陈雄

广东省建筑设计研究院有限公司 首席总建筑师
全国工程勘察设计大师
国务院特殊津贴专家
当代中国百名建筑师
教授级高级建筑师 硕士生导师
国家一级注册建筑师
全国勘察设计同业协会“杰出人物”
建国 70 周年暨第一届中国建筑设计行业管理卓越人物 · 最佳突出贡献奖获得者
全国建设系统先进工作者
广东省五一劳动奖章获得者
广东省建设系统先进工作者

Chen Xiong

Chief Architect, Guangdong Architectural Design and
Research Institute Co., Ltd (GDAD)
Engineering Survey and Design Master of China
Expert Entitled to the State Council Special Allowance
One of the 100 Contemporary Chinese Architects
Professor-level Senior Architect and Graduate Supervisor
National First-class Registered Architect
Outstanding Figure by China Engineering and Consulting Association (CECA)
Recipient of the Most Outstanding Contribution Award of the First Session
of Excellent Management Figure Selection of Chinese Architectural Design
Industry for the 70th Founding Anniversary of the People's Republic of China
Advanced Worker of the National Construction Sector
Recipient of Guangdong May 1st Labor Medal
Advanced Worker of Guangdong Construction Sector

合和建筑观

INTEGRATED AND HARMONIOUS ARCHITECTURE

陈雄 著
Chen Xiong

中国建筑工业出版社
CHINA ARCHITECTURE & BUILDING PRESS

造烛求明，读书求理，
和心求索，营筑求臻。
As one makes candles for light and reads books for truth,
we pursue harmony and excellence in architecture.

序言

FOREWORD

序一　设计的哲学解读

陈雄大师是广东省建筑设计研究院有限公司的首席总建筑师。在他几十年的建筑生涯中，创作了一批优秀的建筑作品，取得了很好的社会效益、经济效益、文化效益和环境效益，他也是我国建筑界具有责任感，同时十分活跃、有影响力的中年建筑师之一。

我和陈总初识于 2010 年，那时正在广州举办一次学术年会，但据陈总说其实早在 1998 年广州白云机场扩建方案的评审会上我们就见过面，那时他们广东省建院参加了投标，但当时我没有注意，以至失之交臂。后来我在 2012 年准备编辑一本《建筑学人剪影》，收录了我拍摄的 227 位建筑界的人物的肖像，陈总自然也包括在内，于是对于他的工作、创作也逐渐注意起来。知道他于 1979~1983 年在华南工学院建筑系学习，之后又师从我国第一代老建筑师林克明先生和郑鹏教授攻读硕士学位，1986 年起进入广东省建院工作至今。在 2016 年第八届全国工程勘察设计大师评选中，被评为全国工程勘察设计大师，成为他们那一批建筑师（出生于 1960 年后）中的佼佼者，还先后获得过“当代中国百名建筑师”“全国勘察设计协会杰出人物”“建国 70 周年暨第一届中国建筑设计行业管理卓越人物最佳突出贡献奖”等一系列荣誉。最近陈总的新作《合和建筑观》即将出版，邀我作序，在研读了他的理论架构和创作项目之后，我正好利用这次学习的机会，写一点自己的体会。

广东是岭南文化的重要传承地，历史悠久，除保存了族群合力、继承中原文化的特点外，一直有敢于创业、勤于交流、务实勤奋的传统，近代以来又成为民主革命的发源地。早在六十年前，我曾在北京参观过岭南派画家陈树人先生的画展，深为岭南艺术界“折衷中西，融合古今”的特点和革新精神所折服。改革开放以后，广东又成为我国开放程度最高、经济活力最强的区域，在科技创新方面充满了活力，其影响力远及全国以至国外。即以建筑界而言，自 20 世纪的 70 年代以来，也是国内建筑界时时前往取经学习的地区。60 年代的北园酒家、泮溪酒家、矿泉客舍等就以不同于北方建筑的独特风格和手法给人耳目一新之感。开放之后自广交会及一系列的对外宾馆更是引领风气之先。广东省建院作为广州地区的主要设计单位，在黄远强总建筑师、郭怡昌大师、容柏生院士、胡镇中总建筑师等带领下，也有一系列优秀作品问世。尤其是 1980 年，由郭怡昌大师领衔设计的北京中国工艺美术馆项目，是广东设计单位在北京长安街上建成的第一栋公共建筑，其高低错落的布局、材料和色彩的运用、传统与现代手法的结合充分体现了岭南“中外共冶一炉”的理念，独具特色与风格，陈总当年也参加了这一工程的设计。岭南文化的背景、开放共存的思路、绿色生态的环境、创造与革新的传统成为陈总本书写作时重要的基础和背景。

建筑设计工作是建筑师所从事的一项创造性的实践活动，是为了改善人们的生存环境、生活条件，向遇到的各种困难、制约而努力设法加以解决，复杂的整合和集成的过程。陈总在长期的工程实践中，尤其是自己所经手的大型和超大型交通建筑、体育建筑、会展建筑、城乡的开发和更新过程中，都要妥善地处理人流、物流、信息流的复杂关系，同时又是高投入、高科技、高风险的大项目。在工作的实践过程中，肯定遇到对许多矛盾和问题的深入追问和思考，其实这就是哲学的思考，就是从实践出发，又要超越实践，从而追求一种更为完美的理想的解决方案，这实际就是实践哲学所要解决的问题。当今像陈总一样的建筑师们不是单纯地从表面的建筑形式问题入手，更多地要从哲学观和方法论的角度和层次，去分析矛盾和解决矛盾，从而把控和处理这些矛盾，才能设计出“好用、好看、好建、好管”的建筑。

陈总所选择的切入点就是本书所集中论述的“合和建筑观”。这是从中国的传统哲学思想中，结合当前的时代特点和潮流所提出的一种解读。由于建筑的服务对象是人，因此“以人为本”是研究问题的基本出发点，是研究人和自然、人和物、人和群体关系如何整体处理。在“合”的层面，我们最常常引用的就是“天人合一”。实际上，古代在探索人和自然的关系中，有着“天人合一”“天人之分”和“与天地参”三种主要学说，分别代表了自然和人为的息息相通、和谐统一；自然和人的区别，用积极的人为来改造自然；强调二者之间既有区别，又有统一的基础上，人可以参与自然界的变化这样三种学说。其代表如孟子的“尽心知性知天”，庄子的“天地与我并生，而万物与我为一”；而荀子则主张“天有其时，地有其财，人有其治”；《中庸》则强调“能尽人之性，则能尽物之兴。能尽物之兴，则可以赞天地之化育，可以赞天地之化育，则可以与天地参矣。” 在这三种不同的对待天人关系的学说中，并不完全是互相排斥和对立，而是在人和自然的关系中，是适应自然，还是改造自然，还是强调人在其中的主观能动性，实际表现了对“合”的辩证统一认识。

而“和”的原意为相应。如孔子的“君子和而不同，小人同而不和”，《国语》中西周末年史伯谓“和实生物，同则不继”，“以他平他谓之和”，主张不同事物间的差异、矛盾及其平衡、统一才能产生新的事物。这些提法最后引申为“和谐”之意，其中心就是积极辩证地看待自然和社会中的差异和矛盾，提倡发挥不同个性各自的积极作用，并在此基础上实现整体的和谐与发展。“合和”的观点体现了中国古代思想家们的原创生存智慧，是思想上的大进步，是对世界文化史的重要贡献。陈总正是由此出发，“传承于传统的哲学理念，发展于岭南地域建筑观念，启发于当代国际建筑思潮，希望涵盖建筑从宏观到微观的多个层面，使要素之间整体‘合和’”。“通过对各因子的有机整合，融入创新理念，最终实现多要素的和谐统一，多维融合，合以和谐。”

在确定了建筑设计的核心观念，从哲学高度来认识建筑设计之后，如果没有与之相适应的方法、策略、手段以表现建筑设计行业的专业特征和本质，也是无法完成这一实践活动的。尤其是建筑设计的创造过程，有着比一般的科技活动更为复杂的关系和内涵，其解决路径和最后选择的解决方案并不是唯一定解，而是要通过集成、综合、协调、权衡，从而在特定的环境、城市、地段、经济、使用、社会、美学等诸多限制条件下，经过选择决策出一个相对优化的方案。人们常常体会一个工程项目从可研、构思、设计方案、施工图到最后完成，对建筑师的团队来说，就是一个从理想主义者过渡到现实主义者的过程。陈总从自己执业多年中以及面对各种类型的大型公共建筑的设计积累的经验出发，在如何处理设计的全过程中从“向好”的目标逐步总结出了自己的八维要素系统构成，即：

功能维度：集约复合、整合共生、水平融合、纵向拓展、规划弹性预留、功能承接重组；
交通维度：立体叠加、通廊设置、流程设计、无感流程、城市路径建构、建筑路径串联；
空间维度：空间组织、公共空间置入、多功能空间与可变空间；
形态维度：多元共存，形态的意向表达和表皮建构；
生态维度：自由呼吸建构、立体绿化嵌入、生态有机适应；
人文维度：文脉元素、交往场所、情景感知；
经济维度：全生命周期成本控制，多重价值的实现；
创新维度：探索创新实践，探索未来发展。

陈总所提出的八维要素与工程方法论中所提出的结构化、功能化、效率化和和谐化的基本特征是相一致的。首先他将建筑设计系统形成了较完整的结构框架，把设计的整体性思维与过程中的结构性思维结合，以逐步形成结构优化的设计体系。其次是处理结构与功能间的辩证关系，不同要求的工程如何审时度势将有关维度加以融合和运用，通过相互结合和促进，以实现设计工作的最终目标；在设计工作的多种维度的动态耦合过程中，实现有序、连续和互相协同的集成，保证提高其效率，其中包括技术效率、经济效率、社会效率和文化效率的实现；最后就是通过设计工程本身的和谐，进一步达到团队内部的和谐、团队与自然、与社会的和谐。有关的维度展开及运用，在陈总的论文和实例中都有不同程度的生动体现和说明，同时也有进一步的拓展和延伸，此处就不赘述了，总之这里凝聚了陈总和他的团队多年耕耘的努力付出和心血。

建筑设计工作的另一个重要特点是，在核心价值观确定以后，其方法和策略并不是一成不变的，其中充满了机动、变化、延伸、拓展的内容，其主要框架结构系统和复杂的子系统，始终处于动态、调整、补充和完善之中。在地域适应性，中国传统“和合”思维滋养，高密度城市化，多元需求和可持续发展的背景之下，“‘合和’倡导多维要素的融合与平衡，实现矛盾与多样性的统一，这些要素既相对独立，又互有影响，彼此联系紧密，形成动态的合和建筑观的理论内核”对设计也有重要的指导意义。

现代建筑设计工程活动的发展和变革，其速度之快、内容之丰富使建筑师应接不暇。2019 年 3 月，中共中央和国务院印发了《粤港澳大湾区规划纲要》，表明在国家发展的大战略中以广州为中心的珠三角九城市在实现新时期新征程的奋斗中所占有的重要地位，是在此基础上，培育新优势，发挥新应用，实现新发展，作出新贡献的大机遇、大文章。为实现这个世界级的城市群，有全球影响力的科技创新中心，“一带一路”建设的重要支撑，内外合作的示范区的重大任务。对于陈总和他们的团队提出了更紧迫和严峻的挑战，“合和建筑观”已经在几十年的实践中取得了丰硕的成果，在新的挑战面前，核心理念还会继续充实、丰富，其设计方法也会在新形势和新要求的大环境下，经受考验，同时发挥其韧性和机动性，不断完善和提升。

预祝年富力强并正处于创作旺盛期的陈总的新书取得成功，也祝他和他的团队在新时期大湾区这样的大舞台上，在他提出的“创造对社会有价值的建筑精品”的目标下，有更多的优秀设计作品问世！

马国馨
中国工程院院士

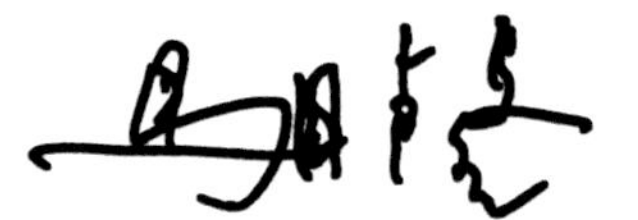

FOREWORD I: A PHILOSOPHICAL INTERPRETATION OF DESIGN

Chen Xiong is a National Engineering Survey and Design Master in China and the chief architect of GDAD. During his decades of architectural practice, he has created many impressive works of satisfactory social, economic, cultural and environmental outputs, which made him one of the responsible and influential middle-aged architects active in the architecture community in China. My first acquaintance with Mr. Chen was in 2010 at an annual academic conference held in Guangzhou. Later in 2012 when I was about to compile Profiles of Architects, a collection of my architect portrait photos, Mr. Chen was naturally included in. It was then that I got to know his work and design. Recently, Mr. Chen invited me to write a foreword for Integrated and Harmonious Architecture, his new book to be published. Availing this opportunity, I was able to write something about my humble opinion.

Guangdong, with a long history, is an important heritage place of Lingnan culture and the birthplace of democratic revolution in modern times. Since China's reform and opening up in 1978, Guangdong has become the most open and economically vibrant region in the country. For Chinese architects, the province has been an important destination to visit and to learn from since the 1970s. GDAD, as a major design institution in the region, has produced numerous successful works as led by Chief Architect Huang Yuanqiang, Design Master Guo Yichang, Academician Rong Baisheng and Chief Architect Hu Zhenzhong. In a word, it is the cultural context in Lingnan region, the open and symbiotic concept, the green and ecological environment, and the tradition of creation and reform that constitute the important foundations and basis for Mr. Chen's book. Architectural design is a creative work practiced by architects. Over his decades of architectural practice, Mr. Chen must have encountered, inquired and explored into many conflicting and problematic issues. Such inquiries and explorations must be rooted in yet go beyond the practice to pursue a better and more ideal solution. Like Mr. Chen, many architects today do not simply tackle the issue from the superficial architectural form, but more from the perspective of philosophy and methodology to analyze and solve conflicts, so as to design buildings that are "easy to use, easy on the eyes, easy to build, and easy to manage".

The perspective Mr. Chen has chosen is the "integrated and harmonious architecture" as explained in this book. It offers an interpretation inspired by the traditional Chinese philosophy and incorporating the characteristics of modern times and trends. In ancient exploration of the relationship between man and nature, there were mainly three theories, namely, "unity of man and heaven", "distinction between man and heaven" and "a trinity of man, heaven and earth". These three theories highlight the subjective initiative of man in respect of whether adapting to or transforming nature, signifying a dialectical and unified understanding of "合(he)", i.e., "integration". The original meaning of "和(he)" was to echo, and was later explained as "harmony", the core of which is to view the differences and conflicts in nature and society in a positive and dialectical manner. The view of "合和(he he)", i.e., "integration and harmony", embodies the original survival wisdom of ancient Chinese thinkers and represents an important contribution to the world cultural history. It is from this perspective that Mr. Chen developed his theory of Integrated and

Harmonious Architecture which is "derived from the traditional philosophical thinking, promoted by the Lingnan-style architectural concept and inspired by the world's architectural trends today, with the purpose of blending all architectural components from macro to micro level into an 'integrated and harmonious' whole", and, "through organic integration of all components and incorporation of innovative concepts, eventually achieve harmony and unity of all".

With a well-established theoretical core and a philosophical understanding of architectural design, Mr. Chen, based on the profound experience he had gained from decades of design practice with a wide range of large public buildings and during the whole design process of "continuous improvement", gradually summarized an eight-dimensional factor system that covers function, traffic, space, form, ecology, humanity, cost, and innovation. This system precisely aligns with the basic characteristics of structure, functionality, efficiency and harmony in engineering methodology, and well represents the efforts and outputs of Mr. Chen and his team in past decades.

Another important aspect of architectural practice is that, even with an established theoretical core, the design approaches and strategies may still vary with projects, and the main framework system and complex subsystems are always dynamically adjusted, supplemented and improved. Against the background of regional adaption, high-density urbanization, diversified modern needs and sustainable development, the Integrated and Harmonious Architecture, as inspired by the traditional Chinese philosophy of integration and harmony, "pursues integration and balance of multi-dimensional factors of architecture to realize unity out of conflicts and diversity, and views these factors as both relatively independent and mutually influential and interrelated in shaping a dynamically developing theoretical core of the Integrated and Harmonious Architecture", which is of great significance in directing the design.

Nowadays when architects are often overwhelmed by the drastic and profound development and evolution in architecture and engineering practice, the Integrated and Harmonious Architecture has seen fruitful results in project practice in the past decades. In face of new challenges, its theoretical core will continue to be enriched and its design approaches will also stand the test of new situations and requirements, meanwhile remain resilient and flexible for further improvement and upgrade.

I wish Mr. Chen, a productive architect in his prime, and his new book a great success. I also look forward to more excellent design works from him and his team in the development of the Greater Bay Area in a new era and on their way towards his goal of "creating more meaningful architectural masterpieces"!

Ma Guoxin

Ma Guoxin
Academician of Chinese Academy of Engineering

序二 根植岭南 营造臻品

我认识陈雄建筑师比较早，我是78届，他是79届，我们都在岭南地域。大家真正熟悉起来是在2013年，当时参加中国建筑学会的中国建筑师代表团一起去欧洲考察，去了匈牙利、捷克和英国三个国家，与这些国家的建筑学会协会进行了学术交流，在这次活动的过程中，我们一直都有交流。陈雄在捷克重点介绍了他带领团队完成的白云机场航站楼和广州亚运馆等项目，给我留下了比较深刻的印象。

在2014年的时候，我参加了广东省建筑设计研究院ADG · 机场院10周年的一个学术论坛和一个展览。全国许多大院的总建筑师都参与了这次活动，在一起进行热烈的分享交流。通过这么一个活动，陈雄所带领的团队也得到了全方位的展示，有系统的作品展览以及创作的总结，还出版了《十年之外 · 十年之间》的团队作品集。应该说，陈雄所带领的团队是一个具有专业性的创作团队，比如说从2004年投入使用的白云T1航站楼，到2018年投入使用的T2航站楼，以及2010年为广州亚运会设计的主场馆等项目，都是一些很出色的建筑作品。很多作品刊登在了《建筑学报》和其他重要的建筑媒体上。陈雄的建筑作品主要分两个方面，一类是大型的交通建筑，另外一类是大型的体育场馆，这两类都属于大型复杂的公共建筑，和医疗建筑一样，也存在着功能的复杂、工艺的复杂、流线的复杂，随着外部环境的经济驱动、制度改变、土地集约、管理变革、万物互联的方方面面的社会进步，这些大型复杂公共建筑也呈现出知识迭代、技术迭代、运维迭代的重要特征，而且这种迭代速度正在变得越来越快。这就需要我们的创作团队能够不断加强外部协作和学习，陈雄所带领的团队可以说是一个学习能力很强的团队。

陈雄在大型复杂的公共建筑的创作的同时，也获得了许多社会和行业的认可。比如说白云机场项目就获得了全国优秀设计金奖、“詹天佑奖”、“全国十大建设科技成就”等荣誉和重要奖项。作为这么大型的机场迁建工程，他们承担了其中的核心项目即T1航站楼的设计，应该说在当时全国机场建设领域，或者说当时在华南地区，是最复杂的大型公共建筑之一，也是我国民航史上一个重要的里程碑，反映了当时中国的建筑技术水平，展示了21世纪初中国大型标志性建筑的先进性与独特性。到后来2010年广州亚运会主场馆——广州亚运馆的国际竞赛，这是一个中国建筑师在重大项目上原创中标并实施的作品，也是团队有代表性的项目，同时这个作品也获得了亚洲建筑师协会荣誉奖和全国优秀工程设计一等奖等一系列荣誉。这两个作品先后刊登在《建筑学报》的封面，体现出行业对这些项目的充分关注和认可。近几年的白云机场T2航站楼、肇庆新区体育中心、珠海横琴保利中心三个项目，也都体现了很好的设计水平和完成度，取得了很好的成绩和荣誉。这些年来，整个建筑行业一直都在尝试和探索新的发展路径，传统大院也在尝试如何做出设计精品，广东省建院在这个发展趋势中承担许多的大型公共建筑项目的设计。陈雄所带领的团队在广东省建筑设计研究院的大平台上，通过中外合作、自主原创，或者是本土联合原创，探索出一条比较成功的创作和技术协作模式，这个也是他们团队是长期以来努力的成果。

这本书结合陈雄本人近40年的一些创作实践，进行了系统的思考和总结，提出了“合和建筑观”理论。应该说理论的提出是基于丰富的建筑设计实践，从哲学观和方法论的角度，提出的富有特色的建筑创作观。

这个创作观我觉得有几个特点，一方面是结合他们作品的在地性进行创作的总结，建筑的地域适应性包括对于岭南气候的回应，对于华南地区的高密高容的城市背景或环境，反映出比较有特色的建筑在地性思考。建筑在地性的本质是建筑回应气候，建筑与环境充分结合。他们团队对城市与建筑的热点问题的探讨及解决方式是比较有特色的，包括集约城市发展，人口不断增加，需要提升土地利用率；探索立体复合空间整合功能综合开发，包括 TOD 开发；在高密度高容积率条件下，进一步提升公共空间的品质；提倡建筑的公共性，与城市空间互动；城市与建筑更加绿色，重点关注节约土地，采取被动节能和主动节能。建筑是文化的重要载体之一，岭南建筑文化是岭南文化的重要组成部分，它必然反映岭南建筑多元兼容，功能优先、不断革新的文化特色。岭南地域的气候与自然对建筑的长期影响，逐步形成建筑的地域风格特征——轻盈、通透、明快，体现岭南地域独特气质。他们团队的作品在建筑造型和空间的在地性语言的探讨也是富有成效的。

第二个方面我觉得是建筑的当代性问题。他们团队的主要作品一直走在行业的前沿，像以机场为代表的交通建筑，应该说是规模很大、功能也非常的复合，系统很多，技术也相对比较先进，设计的手法和实现的路径，也是紧随了国际建筑潮流的发展。他们对于建筑当代性的思考，除了前面谈到建筑对于高密高容城市的回应是一方面，比如说建筑空间构成，根据功能流程用连续动态的空间加以整合连接，从平面到立体，在传统静态空间的基础上突显了当代建筑的流动性；与此相适应的建筑造型语言，也是创新塑造了一系列非线性曲面的形体，以及熟练使用了离散渐变的表皮肌理。创新的造型和空间通过建筑结构一体化的设计得以实现，包括对于复杂建筑造型、复杂大型公共建筑的体系，对于信息模型一体化的实施路径，以及全专业全要素的综合运用，也是他们的一些特点，我觉得他们对于这些方面的思考和实践是富有启发性的，取得了显著的成绩。

第三个方面我觉得是建筑的系统性问题。整个理论的体系，包括理论内核的探讨，包括建筑要素发展的思考，这是多要素的整合和融合，尤其是大型复杂公共建筑所体现的系统性。比如功能、流线、空间、形态等要素，还增加了一些当代发展要素的整合，即绿色、人文、价值和创新这些维度，应该说这两方面的合和，把传统的建筑要素和当代发展的要素整合在一起，运用综合的系统方法去解决复杂建筑的问题，追求“合和”的建筑平衡状态，就此提出了自己的一些创新见解。当然我觉得“合和建筑观”理论的核心还是提倡和谐处理人、建筑和自然的本质关系，这么一个建筑创作观还包括了一些根据他们的实践提出的设计原理，即八维要素的系统构成，同时在八维要素之下，还有相关的一系列设计策略。这是一个贯穿于建筑创作全过程的设计手段，或者设计的语言、手法，提出了他们的富有创造性的观点，并提出了结合他们的实践，能够实施的设计路径，例如复杂大跨建筑的设计数据分析系统，这些都具有一定的先进性。应该这么说，结合丰富的设计实践提出的创作理论，既有相当的理论高度，也具有指导建筑创作的意义。我相信“合和建筑观”也是一个开放的系统，它会结合创作实践不断发展而具有生命力。

我受陈雄的邀请，对于这本包括他的理论和主要作品的专著作序。这是一个非常好的交流机会，可以引起对当代建筑创作以及建筑未来发展的深度思考，期待陈雄大师带领他的设计团队在建筑创作的领域持续大胆创新，继续取得好的成绩，为行业为社会提供更多优秀作品，正如他提出来的“好设计、用心做”的团队核心理念！

孟建民
中国工程院院士
深圳市建筑设计研究总院有限公司
首席总建筑师
深圳大学本原设计研究中心主任

孟建民

FOREWORD II: THE ROAD TO EXCELLENCE IN LINGNAN CONTEXT

I knew Chen Xiong quite early. I am based in Shenzhen and he is in Guangzhou. I was admitted to South China University of Technology (SCUT) in 1978 and he was in 1979. Yet we didn't really get to know each other until 2013, when we visited Europe together as members of a Chinese architect delegation of the Architectural Society of China (ASC). I was quite impressed by the projects of his team he introduced at a presentation in the Czech Republic, such as Guangzhou Baiyun International irport Terminal and Guangzhou Asian Games Gymnasium.

In 2014, I attended an academic forum and exhibition for the 10th anniversary of Guangdong Architectural Design and Research Institute Airport Design Group (ADG), where his team was brought into spotlight, so were their design works and reflections on architectural creation. Indeed, Chen Xiong has led a very competent design team to deliver many successful architectural works, including Guangzhou Baiyun International Airport Terminal 1 & 2, and the gymnasium for the 2010 Guangzhou Asian Games. While focusing on large, complex public buildings, they also worked on other building typologies of distinctive features.

Chen Xiong's architectural practices with large, complex public buildings have been widely recognized by the public and in the industry. For the award-winning Guangzhou Baiyun International Airport project, the Terminal 1 designed by Chen Xiong and his team was considered the largest and most complex public building in airport development in South China and whole country. It made an important milestone in the history of civil aviation in China and showcased the country's visionary and unique design of large landmark buildings in the early 21st century. Later in 2010, Chen Xiong led his team to participate in the international design competition for Guangzhou Asian Games Gymnasium, and won the contract with their original creative design proposal, which was later implemented and won various awards. This was a rare success story for Chinese architect in major projects at that time. The two projects mentioned above both appeared on the cover of the Architectural Journal, well celebrating their passion and efforts for excellence. In recent years, their star projects like Guangzhou Baiyun International Airport Terminal 2, Zhaoqing New Area Sports Center and Poly Center in Hengqin, Zhuhai were quite successful and highly recognized among the architecture community. Through years of hard work, he eventually led his team to identify a development path for large multi-disciplinary design institute to deliver quality design.

This monograph reflects on and summarizes the design practices of Chen Xiong and his team in the past four decades, and proposes the theories of Integrated and Harmonious Architecture, an inspiring architectural creation theory from the perspective of philosophy and methodology.
In my opinion, this architectural creation theory has several distinctive characteristics. The first characteristic is about the site-specific feature of the architecture, which involves consideration of the humid and hot climate in Lingnan region, the high-density and high-FAR urban context, as well as the diverse and inclusive culture. In fact, Chen's team tackle these issues with distinctive and effective solutions, such as intensive urban development, multi-level and composite space, mix-

used functional development, land conservation, passive and active energy conservation, as well as light, transparent and lively architectural style.

The second is about the contemporaneity of architecture. The representative projects of Chen's team remain at the leading position in the industry, while their design approaches and implementation paths also closely follow the global architectural trends. They integrate functional flows with continuous and dynamic spaces to highlight the fluidity of the contemporary architecture. They also create a series of forms with non-linear surfaces and skillfully use discrete and gradually changing skin textures to develop a building form vocabulary that well fits the spaces. They can realize innovative forms and spaces through integrated architecture and structural design, including the implementation path of information model integration, while the comprehensive application of a multi-disciplinary and total factor approach constitutes another feature of the team. Their thoughts and practices in this regard have been quite enlightening and saw fruitful results.

The third is about the systematicness of architecture. The whole theoretical system, including the theoretical core and thoughts on the development of architectural factors, reflects the systematicness of large, complex public buildings. On top of function, circulation, space and form, some contemporary factors such as green concept, culture, value and innovation are also integrated to tackle complex architecture with comprehensive and systematic approaches, leading to the formation of the innovative ideas on the "integrated and harmonious" architectural balance. The core of the theory is to ensure the harmonious relationship between human, architecture and nature, based on which the eight-dimension factor system, design strategies, as well as practice-based feasible design paths are further developed. This architectural creation theory is of great significance, both theoretically and practically. As an open system, it will grow organically along with the architectural creation practice.

Invited by Chen Xiong, I am honored to write the foreword for this monograph of his theory and main works. I look forward to continuous bold innovation and more inspiring works from him and his team, and wish them greater success in embodying the team's core idea of "pursing design excellence with great care and diligence".

Meng Jianmin

Meng Jianmin
Academician of Chinese Academy of Engineering
Chief Architect of Shenzhen General Institute of Architectural Design and Research Co., Ltd.
Director of Benyuan Design and Research Center, Shenzhen University

序三　创新，无尽的前沿

转眼之间，我与机场设计团队一起已经 20 多年了，自己整个职业生涯也近 40 年，在这么一个其实并不短的时间里，自己和团队在建筑创作的道路上不断前行，所有这些成果都来之不易，包含了团队全体员工多年的辛勤劳动、艰辛付出和努力拼搏。正如容柏生院士在团队成立十周年的时候所殷切寄语我们的，要继续持守技术本源，致力设计创新，坚定地走好自己的路。

回想我们八十年代中期刚进广东省建筑设计院的时候，正值改革开放之初，我当时非常有幸跟随到设计院的总建筑师郭怡昌大师，以及设计院老一辈的同事们，跟随他们一起工作。印象特别深刻是郭总多年来一直致力于建筑创作，醉心于设计创新，无论项目的大小，都是呕心沥血地进行创作，进行构思，反复推敲。每有新作，总是以百倍的热情、拼搏的精神和不懈的努力，在创作中施展聪明才华，攀登新的高峰。从郭总的身上我看到了一代建筑大师的风采，很勤奋，很严谨，很睿智但同时他也很谦和，为我们树立了崇高的榜样。后来有机会跟设计院其他的总建筑师胡总、刘总等一起共事，都能够感受到他们的敬业精神，当然还包括对年轻人的关心和培养、指导和提携，我们的前辈给我们一个树立了非常好的榜样。

作为我自己的职业生涯，或者是团队一个重大的机遇，应该是在 1998 年的时候，广州市为新白云机场航站楼举行了国际设计竞赛，我们参加了这次竞赛。虽未中标，但由于体现了较好的技术水平，被业主选定为与国外公司合作共同设计新航站楼，揭开职业生涯和团队发展的重要篇章。这样大型、复杂的建筑项目，在广东省建院历史上也是从来没有做过的，1998 年是一个从零开始的过程。

通过新白云机场航站楼一期工程这样一个高强度的历练，填补了广东省建院在大跨度大空间建筑的一系列关键技术的空白，经过这一次中外合作，我们积累了丰富的大型复杂公共建筑的设计经验，尤其是在大跨度大空间建筑设计的各个相关的专业，广东省建院完成了一个原始的积累，进入了全国建筑设计同行先进的行列。

第二个比较重要的机遇是在 2007 年的时候，参与广州 2010 亚运会唯一的主场馆——广州亚运馆的国际设计竞赛，最终我们团队的方案获得了优胜，这也是中国建筑师少有的在重大项目设计竞赛里原创中标并实施的案例。项目的建设周期很短，只有短短的两年时间，我们从设计、制作到施工配合全过程高完成度地实施了项目。这个项目也是国内最早的大型复杂非线性三维曲面造型的建筑案例之一，以犀牛软件建模，建筑结构共用一套信息模型设计，贯穿方案和施工图设计、工厂制作和施工安装全过程。建筑外形飘逸灵动，成为广州亚运会最为夺目的标志性体育场馆。这个项目开启了广东省建院在大型复杂公共建筑原创设计的新局面。

继白云机场 T1 航站楼之后，我们继续完成了 T2 航站楼设计，项目于 2018 年投入使用，2020 年又开始了 T3 航站楼的设计，为白云机场服务了 20 多年，与业主一起实现了从规划设计到工程实施的“一张蓝图干到底”目标，这在国内外机场领域都很少能够实现的。我们还陆续设计了深圳机场卫星厅，以及佛山新机场、珠海机场、潮汕机场和湛江机场的航站楼，机场领域的设计和技术更新迭代非常快，创新的路上总是你追我赶。

从亚运会到亚青会，我们两次在国际赛事的主场馆原创设计中标，两个项目都面临用地紧张的问题，采用了相同的设计策略去解决复杂功能与空间营造的矛盾。在用地紧张情况下，如果按传统的模式去做，其实很难做到比较好的环境和功能平衡。亚运馆是第一次创造性地把各场馆连在一起，产生了积极的开敞的灰空间，使得四个场馆连成一个群体，在集约的用地下创造了非常好的、高品质的公共空间，这些公共空间是面向赛时和赛后的利用。赛时是集散的地方，有些室外广场，有些室外的灰空间；赛后这些空间又提供给市民高品质的开放空间。近期正在建造的顺德德胜体育中心是这种集约布局模式的最新探索，把场馆的屋面也作为城市开放的公共空间，以更加立体的建构方式解决城市用地紧张和高品质公共空间营造的矛盾，团队的体育建筑也走在原创设计不断创新的路上。

这么多年来，我们以机场航站楼为代表的超大型交通建筑，以及亚运馆为代表的体育场馆这两类项目为主要的业务类型。与此同时，我们还开拓了其他类型的大型公共建筑，包括超高层和开发类的项目，近年来当然也包括一些中小型的项目。无论是合作设计，还是独立原创设计，又或者联合原创设计，我们在完成一系列重要作品的同时，结合业务特点和利用项目平台，策划和进行了面向实施的多种形式的学术研究，包括论文、专著、专利、论坛、评审、授课、出访，在理论上总结提高，开展对外重点宣传，寻求更高的附加值。今天，我们在结合岭南建筑创作不断进行探索的基础上，提出了“合和建筑观”这样一个基于我们这么多年实践和思考的理论。

“合和建筑观”首先还是基于大型复杂公共建筑功能的集约和复合，还有相应的多层次空间，而空间又是由高效的交通组合在一起。复杂的功能和空间，以及交通的联系，映衬在它的建筑形态上，也产生比较多元丰富的建筑语言。还有就是建筑的人文价值，绿色生态有机平衡、建筑创造的价值和成本控制等。这么多年来，我们持续地关注建筑创作的前沿，对于设计的创新一直贯穿在实践中，这也反映出我们整个团队的不断创新的理念。以大型航站楼为代表的交通建筑是多学科融合和创新的典型例子，其规模宏大、功能复杂、系统繁多，技术比较先进，需要建筑师带领设计团队面对复杂的系统与各相关使用方的需求，还有和专业的顾问团队进行密切的配合，融合众多学科和复杂的技术体系，在这个基础上进行大力的创新，通过一体化的设计，才能实现高水平的设计和建设的目标。

“合和建筑观”在设计原理及其主要的核心要素包含了很多创新点，我们希望能够把它们相对独立分开，但互相又是一个紧密联系的整体，作为产生优秀建筑作品的工作流程而进行总结和提炼，以指导我们未来的实践，这是我们进行理论研究很重要的目的。回过头看，几十年的建筑创作实践，我和团队之间的配合、互相促进的学习，都是非常有意义的。在这本著作里包括理论的思考、设计实践总结，包括实施路径以及主要的设计成果。它并不是普通意义上的作品集，我觉得它更多的是通过这些作品展现了创新的理念和创作的过程。我们遇到了一些什么问题？我们怎么去解决这些问题？在这个过程里我们怎么寻求创新点？这些都是我们一直持续追求的。建筑师是长期积累的一个职业过程，通过与团队的合作模式，来实现这些大型复杂的项目。因此我觉得我们这么多年能够设计和建成一批好的项目，完全有赖于我们有这么一些优秀的团队的持续努力。这么多年我们培育了很好的团队，这是我们非常值得欣慰的。我们为团队的同事们创造了一个比较好的氛围，使得他们能够在设计中得到充分的发挥，有一个比较好的成长空间，我觉得这个非常重要。近年，与刚成立的城市工作室在创作上也展开了多项目合作，鼓励年轻的同事追求探索。

我们在国际化浪潮中探索了“合作 - 自立”的发展模式，从 1998 年的新白云机场一期航站楼开始，与国际事务所深度合作，到 2008 年广州亚运馆原创中标实施，继而完成一批原创设计项目，并继续与国际团队深度合作，在“原创设计 + 高端合作”的轨道上发展。实现了从合作设计 - 主创设计 - 原创设计的飞跃，开创了国有设计机构改革发展的新模式。

面向未来，我觉得我们要以中国一流的设计企业作为发展的目标，持守技术本源，致力设计创新，坚定地走好自己的路，在快速发展变化的社会经济大环境下面，我觉得我们必然要走一条持续创新的道路。在设计理念上创新，在技术科研上创新，在开拓经营上创新，在运营管理上创新，在团队的建设上创新。应该说在创新的道路上，其实还有很多事情去做。创新是一种精神，是一种姿态，是无尽的前沿！

从事建筑设计行业的建筑师们和工程师们是幸运的，我们的作品可以长久地存留大地。正因为如此，我们的责任也是异常重大的，我们想象中的世界构成了现实世界的一部分，建筑是一个关系到未来世世代代的重大责任，我们确实任重道远！

FOREWORD III: INNOVATION, AN ENDLESS FRONTIER

I've been working with the airport design team for more than 20 years, and I myself have engaged in this professional career for nearly 40 years. Nothing fruitful ever comes easy, or could have been achieved without years of hard work of all the team members. As Academician Rong Baisheng earnestly expected in the 10th anniversary of the founding of the team, we should stay focused on technology, remain committed to design and innovation, and forge ahead unswervingly.

In the mid-1980s, when I joined GDAD, I was fortunate to work under the guidance of architectural design master Guo Yichang, the then GDAD chief architect who remained committed to architectural design and design innovation. In him, I was able to find something that truly makes an architectural design master of our generation. Later, I also had the opportunity to work with other chief architects of GDAD, such as Mr. Hu and Mr. Liu, whose professional dedication truly impressed me, as well as their care, nurture, guidance and support for young people.

In 1998, we participated in the international design competition for the new Baiyun Airport Terminal in Guangzhou, which was a major opportunity for both my career and my team. We did not win the competition, but our high level of expertise impressed the client, who then designated us to design the new terminal in collaboration with an overseas design firm. Through design practice and collaboration with foreign architects in Phase I of the new Baiyun Airport Terminal, GDAD was able to gain experience in a series of key technologies for large-span and large-space architecture, and became one of the leading architectural design firms in China.

In 2007, our participation in the international design competition for Guangzhou Asian Games Gymnasium, the only main venue for the Guangzhou 2010 Asian Games, offered us another major opportunity. Our success in the design competition was a rare case, as seldom was Chinese architects able to win and implement their original design in major design competitions. This project, as one of the earliest architectural design cases featuring large-scale, complex, non-linear three-dimensional curved surface in China, also marked GDAD's new chapter of original design for large-scale complex public buildings.

After completing the design for Baiyun Terminal 1, we continued with Terminal 2 and then Terminal 3. Serving Baiyun Airport for more than 20 years, we have, working with the client, realized the goal of "following one blueprint throughout the whole project" from planning and design to project implementation. We have also designed a number of other terminals, such as the Shenzhen Airport Satellite Concourse and Foshan New Airport. Airport design and technologies are updating so fast that we must stay competitive through sustained innovation.

From Asian Games to Asian Youth Games, we have won twice in international competitions with our original creative design proposal of main venues. In response to the limited land available for both projects, we adopted the same design strategy to tackle the conflict between complex functions and spatial constraints. For Asian Games Gymnasium, we, for the first time, creatively connected all the venues together, creating positive open grey spaces. The recently on-going Desheng Sports Center in Shunde, with roof designed as an open urban public space, also reflects our latest exploration of this intensive layout model, and the team's continuous innovative efforts in creating original design for sports buildings.

Over the years, we have been focusing on the designs of super large transportation buildings represented by airport terminals and sports venues represented by the Asian Games Gymnasium. Meanwhile, we have also designed other types of large public buildings, including some small and medium-sized projects in recent years. In view of our business focus and capitalizing on the project platform, we have planned and launched various forms of academic research oriented towards implementation. Today, based on our decades of exploration, practice and thinking of Lingnan architecture, we put forward the theory of "Integrated and Harmonious Architecture".

"Integrated and Harmonious Architecture", above all, is based on intensive and integrated functionality of large and complex public buildings, as well as on the corresponding multi-level spaces connected via efficient circulation. Architectural form, therefore, often reflects complex functions and space, and traffic connections, as well as the humanistic value of architecture, green, ecological and organic balance, value of architectural creation and cost control. Over the years, we have never stopped implementing our design innovations in our project and practice. Transportation architecture represented by large terminals is a typical example of our multidisciplinary integration and innovation, which requires the architect and the design team to respond to complex systems and needs, cooperate closely with speciality consultants, integrate numerous disciplines and complex technical systems, and on this basis, carry out vigorous innovation and integrated design to achieve high-level design and construction goals.

The design principles and core elements of "Integrated and Harmonious Architecture" are innovative in many aspects. They are relatively independent individually yet closely integrated as a whole to serve as the workflow for creating excellent architectural works. This book includes theoretical thinking, design practice summary, implementation approaches and the major projects realized. It is the sustained efforts of our teams that enabled us to successfully design and deliver these projects in past years.In recent years, we also worked with the newly established UAD Studio on many projects, encouraging young designers to pursue their careers through exploration.

In the trend of globalization, we explored the development mode of "collaborative design - independent design", developed along the path of "original design + high-end collaboration", and realized the leap from collaborative design - lead design - original design, establishing a new reform and development mode for state-owned design institutions. In the future, we will strive towards the goal of ranking among China's first-class design enterprises, through innovation in design concept, technical research, business expansion, operation and management, and team building. As we believe, innovation is a spirit, an attitude, and an endless frontier!

Architects and engineers can produce works that last for long, so our responsibility is heavy and our tasks arduous, as the world we conceive will constitute part of the real world for generations to come.

Chen Xiong

目 录

CONTENTS

结缘建筑 扎根岭南

MY CAREER AS AN ARCHITECT IN LINGNAN REGION

结缘建筑　扎根岭南

我于 1986 年华南工学院（现华南理工大学）硕士研究生毕业后进入广东省建筑设计研究院（现广东省建筑设计研究院有限公司，以下简称广东省建院）工作至今已近 40 载，扎根于岭南地区，结合大型复杂公共建筑工程的设计实践，不断进行探索研究。

专业的“月老”与“引路人”

听从父亲的建议，我于 1979 年进入华南工学院（现华南理工大学）建筑学系学习建筑学，时逢改革开放拨乱反正，建筑设计还鲜为人知。父亲 20 世纪 50 年代去过香港建筑师事务所，目睹建筑师趴图板画图，而我小时候也喜欢画画、做模型，父亲觉得我有这个基础，因此建议我去读建筑学，有机会为祖国建设出力。

踏入建筑系的门槛，在一众老师的悉心指导下，才知道建筑设计的原理。从“数理化”思维向设计思维转型，过程并不轻松。大学毕业后，我继续在华南工学院建筑设计研究院攻读硕士研究生，有幸师从林克明先生和郑鹏先生，两位导师鼓励创新，将教学与实践相结合，提倡对建筑创作中传统与现代相结合的辩证思考，他们严谨的治学态度，对建筑创作的执着精神，给了我极好的指引，也使我对岭南建筑有了更深入的理解。

深耕近 40 载

1986 年硕士研究生毕业后，进入广东省建院（现广东省建筑设计研究院有限公司）工作。1987 年至 1990 年，有幸跟随省院总建筑师郭怡昌设计大师设计了北京的中国工艺美术馆、陶然宾馆，以及东莞理工学院等项目，在大师身边耳濡目染，常常被他精巧的构思折服， 更敬重他的高尚人品和执着坚忍的创作精神，深深地懂得建筑精品的产生离不开艰苦的建筑创作，只有付出辛勤的劳动才可能取得丰硕的成果。

1998 年有一个契机，广州白云机场 T1 航站楼要建设，我参与了设计。新白云机场一期工程为广东省建院跟国外公司合作设计，在此过程中，我们对于大跨度、大空间，特别是机场这种类型建筑进行了一个全面系统的技术研究。2007 年，广州举行广州亚运馆国际设计竞赛，很长一段时间以来，中国大型公共建筑的国际竞赛绝大部分中标者都是境外公司。综合之前的大跨度技术体系和成熟经验，2008 年 5 月，我们的方案以中国原创获胜并且成为实施方案。近年来，我们通过“三师下乡”志愿服务保持对广阔的乡村区域，特别是贫困乡村的关注，希望以专业技术回馈社会，设计了南粤古道梅岭驿站等众多对社会有价值的小建筑。“好设计、用心做”始终作为我们团队的重要理念，在岭南建筑设计国际化浪潮中探索“合作—自立”的发展模式，对不同类型、规模的建筑进行精细化设计，创造对社会有价值的建筑精品。

乘着改革开放的巨轮，在近 40 载的职业生涯中，从跟随前辈建筑师到自己带领团队，完成了包括交通建筑、体育建筑、会展建筑、开发类建筑、城市更新与乡村振兴等众多类型建筑作品，由珠三角辐射粤东、粤西、粤北，跨越了岭南的大部分地区。在岭南建筑设计实践中密切回应城市，紧密联系时代，持续追求品质。未来，还将持守技术本源，致力设计创新，走好建筑师的路！

陈雄（左二）和师姐、师兄与林克明先生合影
Chen Xiong (back row, second from left) with Mr. Lin Keming and other architecture majors

陈雄（前排左一）与同学一起熬夜画图
Chen Xiong (fornt row, first from left) and his classmates working on design drawings till late at night

中外合作团队留影（后排右二为陈雄）
Chen Xiong (back row, second from right) and colleagues on an overseas business visit

陈雄参加广州新白云机场首航（二排右一拿花者为陈雄）
Chen Xiong at the inaugural flight of the new Guangzhou Baiyun Airport (the first person on the right of the second row, with a bouquet in his hands)

广州白云机场 T1 航站楼
T1, Guangzhou Baiyun International Airport

主要作品分布图
Distribution map of main works

广东	Guangdong
广州	**Guangzhou**
广州白云国际机场一号航站楼	Terminal 1, Guangzhou Baiyun International Airport, Guangzhou
广州白云国际机场二号航站楼	Terminal 2, Guangzhou Baiyun International Airport, Guangzhou
广州白云国际机场三号航站楼	Terminal 3, Guangzhou Baiyun International Airport, Guangzhou
广州亚运馆	Guanghzou Asian Games Gymnasium, Guangzhou
广州花都东风体育馆	Dongfeng Gymnasium, Huadu District, Guangzhou
广州空港商务区会展中心	Guangzhou Aerotropolis Development District (GADD) CBD Exhibition Center, Guangzhou
广州南站核心区 TOD（B 区）	TOD of Guangzhou South Railway Station Core Area (Zone B), Guangzhou
广州猎桥桥西 110kV 变电站	110kV Lieqiao Substation, Guangzhou
广州东照大厦	Dongzhao Building, Guangzhou
广州白云山柯子岭门岗	Baiyun Mt. Entrance (Keziling), Guangzhou
广州明珠湾起步区域市总师城市风貌管控	Urban Landscape Control of Guangzhou Mingzhu Bay Kick-off Zone under Chief Urban Designer Mechanism, Guangzhou
广州荔湾海南经济联社更新城市设计	Urban Design for Renewal of Hainan Economic Association, Liwan, Guangzhou
广州荔湾海北村更新片区城市设计	Urban Design for Renewal of Haibei Village, Liwan, Guangzhou
深圳	**Shenzhen**
深圳宝安国际机场卫星厅	Satellite Hall, Shenzhen Bao'an International Airport, Shenzhen
深圳宝安国际机场新航站区交通中心	GTC, New Terminal Area, Shenzhen Bao'an International Airport, Shenzhen
深圳机场东综合交通枢纽概念方案	Concept Design of Shenzhen Airport East Integrated Transport Hub, Shenzhen
深圳机场 T4 片区规划及航站楼方案	Design Proposal of Terminal 4 and Terminal Area, Shenzhen Bao'an International Airport, Shenzhen
佛山	**Foshan**
珠三角枢纽机场（佛山新机场）方案	Design Proposal of The Pearl River Delta Hub Airport (Foshan New Airport), Foshan
顺德德胜体育中心	Shunde Desheng Sports Center, Foshan
广东（潭洲）国际会展中心	Guangdong (Tanzhou) International Convention and Exhibition Center, Foshan
香港新福港地产 · 佛山新福港广场	SFK Properties · SFK Plaza, Foshan
东莞	**Dongguan**
东莞中心区海德广场	Dongguan Commercial Center Zone F (Haide Plaza), Dongguan
东莞理工学院（参与设计）	Dongguan University of Technology, Dongguan(as an architect)
珠海	**Zhuhai**
珠海金湾国际机场 T2 航站楼	Terminal 2, Zhuhai Jinwan International Airport, Zhuhai
第十三届中国航展扩建工程	The 13th Airshow China Expansion Project, Zhuhai
珠海横琴保利中心	The Poly Center in Hengqin, Zhuhai
珠海国际科技创新中心	Zhuhai International Science and Technology Innovation Center, Zhuhai
珠海横琴科学城（二期）标段二	Hengqin Science City (Phase II) Bid Section II, Zhuhai
汕头	**Shantou**
汕头亚青会场馆	Shantou AYG Stadium, Shantou
韶关	**Shaoguan**
南粤古驿道梅岭驿站	Meiling Station of South China Historical Trail, Shaoguan
湛江	**Zhanjiang**
湛江吴川国际机场航站楼	Wuchuan International Airport Terminal, Zhanjiangl
肇庆	**Zhaoqing**
肇庆东站交通换乘枢纽	Zhaoqing East Railway Station Transportation Hub, Zhaoqing
肇庆新区体育中心	Zhaoqing New Area Sports Center, Zhaoqing
肇庆新区喜来登酒店	Hotel Sheraton Zhaoqing Dinghu, Zhaoqing
惠州	**Huizhou**
惠州金山湖游泳跳水馆	Jinshan Lake Swimming and Diving Natatorium, Huizhou
揭阳	**Jieyang**
揭阳潮汕国际机场航站楼	Jieyang Chaoshan International Airport Terminal, Jieyang
揭阳潮汕机场航站区扩建	Terminal Aiea Expansion, Jieyang Chaoshan International Airport, Jieyang
北京	**Beijing**
中国工艺美术馆（参与设计）	China National Crafts & Arts Museum, Beijing(as an architect)
陶然宾馆（参与设计）	Taoran Hotel, Beijing(as an architect)
武汉	**Wuhan**
武汉火车站（参与设计）	Wuhan Railway Station, Wuhan(as an architect)

MY CAREER AS AN ARCHITECT IN LINGNAN REGION

Nearly four decades have elapsed since I joined the Guangdong Architectural Design and Research Institute (now Guangdong Architectural Design and Research Institute Co., Ltd., GDAD) after I graduated from South China Institute of Technology (now South China University of Technology, SCUT) in 1986. Based in Lingnan Region ever since, I've kept on exploring and studying in the design practices of large-scale complex public buildings.

Predecessors in My Career

Following my father's advice, I became an architecture major in the Department of Architecture of South China Institute of Technology (now SCUT) in 1979, when architectural design was still little known at the time and the reform and opening up just started. My father visited an architect's office in Hong Kong in the 1950s and saw how architects worked on their drawing boards. When I was a kid, I liked drawing and making models. My father saw my potential and suggested that I pursuit a career in architecture and contribute to the development of the country.

After I started to study in the Department of Architecture and under the careful guidance of several excellent teachers, I began to understand some principles of architectural design. The transition from STEM thinking to design thinking was never easy. After four-year study of bachelor degree courses, I continued to pursue a master's degree at SCUT Architectural Design and Research Institute, and had the privilege of studying under Mr. Lin Keming and Mr. Zheng Peng. Mr. Lin and Mr. Zheng combined teaching with practice and encouraged dialectical thinking about traditional and modern architecture. Their rigorous academic attitude and dedication to architectural creation have always been the guidance during my career and helped me develop a deeper understanding of Lingnan architecture.

Nearly Four Decades of Professional Practice

After graduating with a master's degree in 1986, I joined GDAD and from 1987 to 1990, I had the honor to work with Mr. Guo Yichang, chief architect of GDAD in his team in design of China Arts and Crafts Museum in Beijing, Taoran Hotel, Dongguan Institute of Technology and many other projects. I was often impressed by his ingenious ideas, and highly esteemed his noble character and perseverance in architectural design. I started to understand that great masterpieces come from arduous efforts and success can only be achieved by working hard.

In 1998, Guangzhou decided to build T1 of Baiyun Airport and I had the opportunity to take park in the design. The Phase I of the new Baiyun Airport was designed by GDAD in cooperation with a foreign design office. During this process, we conducted a comprehensive and systematic technical research on architecture typology with large span and large space, especially airport terminal. In 2007, Guangzhou held the International Architectural Design Competition for the Asian Games Gymnasium. Although for a long time, most winners of international design competitions for large-scale complex public buildings in China were overseas architects, we as Chinese architects won this competition in May 2008 with our original design scheme, which was subsequently detailed for implementation, thanks to our experience and expertise with large span structures.

In recent years, we paid special attention to rural revitalization, especially those poverty-stricken villages through the volunteer service of "bringing planners, architects and engineers to countryside", hoping to return the society with our expertise and designed the Meiling Station of South China Historical Trials and many other small-sized public buildings with significant social value. "Good design and hard work" has always been an important concept of our team. We have been exploring the development mode of "independent design through cooperation with peers overseas" in making the Lingnan architecture global, insisting on refined design for different types and sizes of buildings, and creating elaborate architectural works that are valuable to the society.

Growing along with the development of the reform and opening up in my career of the past four decades, I have been worked together with predecessor architects and led my own design team to complete a diverse range of projects, from transportation hubs, sports buildings, exhibition venues, development zones to urban renewal and rural revitalization and from the Pearl River Delta to the eastern, western, and northern Guangdong, covering most parts of Lingnan region. In the design practice of Lingnan architecture, we always respond to the needs of the cities and the times, and keep pursuing excellence in quality. In the future, we will remain true to our original aspiration about technology, pursue innovation in design and make our due contribution as architects.

CHAPTER I

合和建筑观理论综述

INTEGRATED AND HARMONIOUS ARCHITECTURE: THEORETICAL REVIEW

1.1 合和建筑观的提出背景

伴随着改革开放的进程，中国的城市建设经历了从小城镇到大城市发展的重心变化。岭南地区作为对外开放和城市化建设的排头兵，催生了大量建筑实践，这些实践浓缩了中国 40 多年来翻天覆地的城市化过程，至今形成了高密度的粤港澳大湾区都市圈。在这一进程中，建筑的地域适应性、“和合”思维延续、高密度城市化、当代多元需求以及可持续发展五个维度成为建筑观形成的主要背景，通过我和团队在这几个方面的实践与思考，尝试提出“合和建筑观”理论。

1.1.1 建筑的地域适应性

建筑受地理、历史、文化等因素的长期影响，逐步形成了相对固定的地域风格特征，体现出独特的地域气质。炎热、潮湿、多雨的亚热带季风气候造就了岭南地区炎热潮湿、夏长冬暖、暴雨常见、台风频繁的气候特点。因此，岭南建筑的气候适应性策略，必须考虑隔热、遮阳、通风、防雨、防潮和防风等问题。同时，岭南自然地理形态复杂，有海岸、河流、山地、丘陵、台地及平原，建筑需要结合不同地形地貌进行创作。全球化进程中的当代中国建筑应该兼具本土精神与时代特色，并在当代的建筑实践中保持地方性语言的表达。建筑应结合气候和环境特征，体现地域气质，还应具有源于功能的标志性特色。建筑师应考虑如何集约用地资源和复合利用空间，建筑创作应该积极回应城市需要，持续追求品质，不断融入新技术，面向可持续发展。在实践中思考建筑创作与地域的关系，致力于当代岭南建筑精髓的再诠释。

1.1.2 中国传统“和合”思维的滋养

“和合”思维是中国传统文化中的一种典型思维。《国语》中周太史史伯的“和实生物，同则不继”表现出了古人对“和谐”的认识，认为其是事物之本与天地法则[1]，这一思想经儒道二家进一步发扬，逐渐成为中国传统文化的基本精神之一。中国传统营造也反映出了这种精神的影响，不论建筑中阴阳关系的体现，或形制、体量、装饰、群体的布局，体现出整体和谐的朴素哲学观念，助生了中国传统建筑的形象。这一传统思维延续至今，促成了中国人追求平衡、和谐的当代思想，在一些建筑创作中也有所体现。

1.1.3 高密度城市与高品质建筑营造

岭南地区位于中国城市化发展的前沿，产生了众多高密度城市群，建筑与人口密度大，这既给城市生活带来了丰富活力，同时也使公共资源预留及分配相对不均衡的问题较为突出。在这种背景下，建筑师需要思考如何打造高品质的建筑和城市空间，弥补资源不足的问题，提高城市生活便利程度，促进人与建筑的良性互动，激发城市的有机活力。对于中国城市现阶段的建筑设计，我们不禁重新思考，建筑应该以什么姿态出现于这种“高密度”的城市环境中，并呈现出高品质的结果，才是合适的。

1.1.4 当代多元需求与建筑 ID 演化

“ID”是英文“identity”的缩写，指事物的身份，建筑作为人类社会重要的物质载体，其不同个体拥有着不同的建筑 ID，且受时代需求影响而表现出明显差异，建筑类型与建筑标准一直随时代发展而变化。维特鲁威提出了经典的“坚固、实用、美观”的建筑标准[2]，建筑类型都呈现着兼具功能性与艺术性的特征，受此影响，中华人民共和国成立早期也将“实用、经济、可能条件下注意美观”作为国内建筑活动的方针[3]，体现出建筑 ID 简约朴素的实用主义特征。如今人们生活水平不断提高，对建筑的需求也在同步提升，当代人们已不再满足于单一内涵的建筑 ID。2016 年中国在新的建筑标准中加入了“绿色”的维度，反映出近年社会对可持续发展的关注；此外，人们对建筑体验的需求也越来越多元，建筑的人文属性不断增强；随着行业不断往细分，建筑功能也产生了多样化的发展维度。传统的建筑 ID 已经难以满足时代，需要更多元的建筑定位、更动态的要素维度，与时代相匹配。

1.1.5 建筑的可持续发展

建筑的可持续发展是当今备受人们关注的问题。大城市的发展相应带来了大量社会资源的消耗，自 20 世纪 70 年代石油危机以来，人们愈发认识到这一点，究其原因，能源消耗不仅是关乎国计民生的经济性问题，更是关乎人类未来的生存问题。当下的建筑实践耗能巨大，是社会可持续发展问题中亟待解决的重中之重，通过对现有实践情况的反思，我们可以意识到，从建筑理念、策略、技术到工法等众多建筑实践的环节均具有继续优化的条件，潜藏着可持续发展的可能。中国建筑实践庞大而迅速，随之而来的污染带来了较大的社会环境治理成本，如今国家已将环境保护作为重要国策，各项政策和行业引导已体现出了国人对保护未来人居环境的重视。因为中国人口众多，资源紧张，建筑实践的负面影响更易产生规模效应，需要更加关注建筑的可持续发展。

建筑实践与理论伴随着城市化进程同步发展，建筑的评价标准正变得更为多元。不论是建筑的地域适应性条件、高密度城市下的建筑品质诉求，还是多元需求下的建筑 ID 演化规律和社会可持续发展的推动，都对当代建筑实践提出了更高要求，需要寻求具有持续生命力的建筑理论。

1.2 合和建筑观

1.2.1 建筑本质和建筑要素的发展

建筑理论作为人类建筑实践的总结，可对建筑发展与建筑实践工作提供指引。为提出与建筑实践相适应的理论，需要对当代建筑实践的特征进行理性分析，而这源于对建筑本质和建筑要素发展的思考。

1）建筑的本质——人类在自然中的庇护所

建筑的本质是什么？从历史上看，建筑起源于人们遮风避雨的基础需求。人在从洞穴中和树上来到地面时，当务之急是寻找到一处安全的居所，以抵御自然气候的侵扰和野兽的威胁，于是有了屋顶、墙面、门洞，诞生了最初的建筑。早期建筑虽然较为简陋，但已具备休息、储物和一定的养殖功能，并且能方便人的使用，一些基本的建筑要素已经出现。随着人类社会的发展，人的需求逐渐多样，建筑的内涵在社会价值观影响下也在发生变化，但其核心一直围绕着人与自然的关系处理上。所以，建筑是为了处理这种关系而存在的物质载体，是联系两者的纽带。

2）建筑要素的发展

随着人类文明的发展，建筑被逐渐赋予了更丰富的内涵，围绕人与自然的本质关系，扩展出众多建筑要素。从古到今不同的文明主体曾诞生了不同的独特建筑内涵，古希腊基于毕达哥拉斯数理美学的几何造型，中国基于山水观的群体形象，不一而足，意味着不同地区的建筑观应当有所差别。时至今日，建筑走向了全球化融合与地方特征相互发展的道路，除了功能、流线、空间、形态等传统建筑要素外，出于对环境保护、人本主义、可持续发展的关注，还形成了建筑在生态、人文、经济要素方面的共识。

1.2.2 合和建筑观的理论内核与理论体系

1）理论内核——追求“合和”的建筑平衡状态

合和建筑观的核心观念包括“合”与“和”两个方面。“合”即“整合、融合”，意指统筹建筑的广泛要素，包括对功能、流线、空间、形态等建筑核心要素，和绿色、人文、价值、创新等当代发展要素的整合；“和”有“和谐、中和”之意，指在建筑要素相互联系的基础上，进一步彼此协调形成有机体。这一内核借鉴了中国“天人合一”与“和合”的传统哲学思想，结合岭南建筑兼容并蓄、持续创新的实践精神和当代国际建筑思潮多元开放语境的启发，以实现建筑要素之间的和谐，形成建筑系统整体关系平衡稳定的“合和”状态。在建筑设计中，通过对各因子的有机整合，融入创新理念，最终实现多要素的和谐统一，多维融合，合以和谐，“合和”倡导多维要素的融合与平衡，实现矛盾与多样性的统一，这些要素既相对独立，又互有影响，彼此联系紧密，形成动态发展的理论圈层网络，围绕“人、建筑、自然”的本质关系，共同组成了合和建筑观的理论内核。

2）设计原理——八维要素系统构成

基于合和建筑观的理论内核，针对建筑的内外要素，结合多年实践经验，提出了功能集约复合与弹性可拓、高效交通连接与漫游路径植入、营造空间活力与韧性、多元建筑形态与表皮建构、生态融合与有机平衡、人文介入与体验感知、成本控制与价值驱动、前沿探索与创新实践等 8 个维度的设计原理。功能集约复合与弹性可拓从功能的角度出发，探索建筑功能复合与可持续发展的可能；高效交通连接与漫游路径植入通过研究交通流线、室内外通行流程与人性化体验路径，力求提升交通效率、丰富通行体验；营造空间活力与韧性通过对空间的探索，赋予了具有城市公共功能的现代空间类型与空间内涵；多元建筑形态与表皮建构则以对线性、非线性建筑形态的建构，营造一种符合时代特征，与环境、文化相适应的建筑形态；生态融合与有机平衡强调建筑与生态环境的交融、和谐共生，并通过相关技术手段实现建筑的绿色低碳；人文介入与体验感知通过文

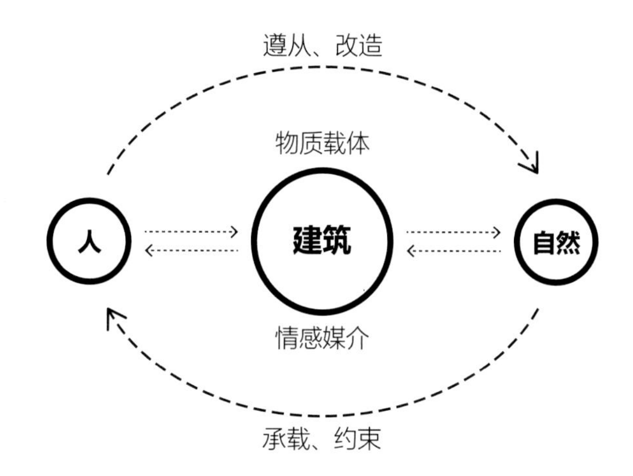

建筑的本质及其与人和自然的关系
The nature of architecture and its relationship with man and nature

脉探寻、符号转译、体验提升、人本关怀，打造具有人性化的场所；成本控制与价值驱动则从全生命周期的建筑成本控制出发，实现社会、经济、艺术、科学的多重建筑价值；前沿探索与创新实践则出于对持续创新的追求，以一种积极的建筑设计态度，努力拓展行业边界，实现建筑的未来展望。值得强调的是，这些设计原理并非一成不变，其具有包容性与开放性，如建筑要素一样，随着时代变化，将处于不断发展之中。

3）设计策略——全过程设计手段

设计策略是基于设计原理进一步提出的具体手段，为近 40 年建筑实践的设计经验总结。在功能维度，提出了集约复合、整合共生、水平融合、纵向拓展、规划弹性预留、功能承接重组等策略，以营造具有复合性、符合城市发展特征的建筑综合体；在交通维度，提出了立体叠加、通廊设置、流程设计、无感流程、城市路径建构、建筑路径串联等策略，以营造通行便捷、体验流畅的建筑交通系统网络；在空间维度，提出了空间组织、公共空间置入、多功能空间与可变空间等策略，以营造自由共享、具备未来发展潜力的空间；在形态维度，提出了多元共存的建筑形态、建筑形态的表皮建构、建筑形态的意向表达等策略，以营造具有时代特征、适应当地的建筑造型；在生态维度，提出了自由呼吸建构、立体绿化嵌入、生态有机适应等策略，以营造兼具环境融合、微气候特征、绿色被动技术使用的绿色建筑体系；在人文维度，提出了文脉元素、交往场所、情景感知等策略，以营造具有人文特色和人性化体验的建筑场所；在经济维度，提出了全生命周期成本控制、实现社会、经济、艺术、科学等多重价值等策略，以营造具有高效可持续的建筑设计与运维体系；在创新维度，提出了探索创新实践的策略，并对未来发展模式进行展望，探索未来的建筑道路。

4）实现路径——高品质建筑实践工具

实现路径是合和建筑观的实践工具，包含创新当代复杂条件下的设计路径、复杂大跨建筑设计数据分析系统、建立互馈闭环的全周期全要素设计流程三个方面。设计路径旨在对项目特征、行业先进理念、先进技术手段等方面提出优化路径，包括多专业交叉融合、全专业多层次的绿色设计、全生命周期与全专业 BIM 协同等方面；分析体系旨在设计中引入理性的数据分析和建筑模拟，确保实现建筑的设想目标，包括基于可量化的建筑空间分析、基于立体多维的建筑环境分析、基于仿真模拟的流程分析等方面；设计流程旨在关注建筑从设计到运营的全过程，确保环节可控，包括策划与设计、施工协调、运维及后评估等方面。实现路径是基于当代建筑实践的经验总结，能适应当代建筑行业的发展，不仅是建筑实践的重要技术要求，更是使建筑品质持续提升的内在保障。

1.2.3 合和建筑观的理论特征

作为一种建筑设计理论，合和建筑观被寄希望于体现出三个特征：一是兼具理论性与操作性。合和建筑观应不仅提出建筑理论的概念，具有核心的价值要义，而且提出建筑设计方法与建筑实践的途径，可以用于指导实践。二是具有一种整体观。合和建筑观传承于传统的哲学理念、发展于岭南地域建筑观念、启发于当代国际建筑思潮，希望涵盖建筑从宏观到微观的多个层面，使要素之间整体“合和”。三是寻求随时代发展的动态平衡。理论应具有开放性与可拓展性，能适应未来建筑设计的持续发展，在动态发展中寻求“合和”的平衡状态。

回顾近 40 年的个人实践，本人主要进行了交通建筑、体育建筑、会展建筑等类型的探索，同时也尝试了一些其他的建筑类型。这些实践对象多为大型公共建筑，具有复杂功能和较强的公共性，服务了大量人群，还是建构城市形象的重要部分，这要求一种多维融合、综合平衡的设计，来满足复杂多元的需求；在近年的小尺度建筑实践中，进一步探索了这种和谐、平衡的设计。通过对当代条件的分析、对经典建筑理念的学习与发展，结合建筑实践的思考提出该理论，以理论内核、原理、策略、路径作为框架，并以具体的实践作为印证，试图表达对建筑本质和要素动态发展的认识，以回应建筑创作的核心问题。希望该理论能在丰富的建筑理论海洋中贡献一份力，并在未来的实践中，不断发展和完善，以展现出持续的生命力。

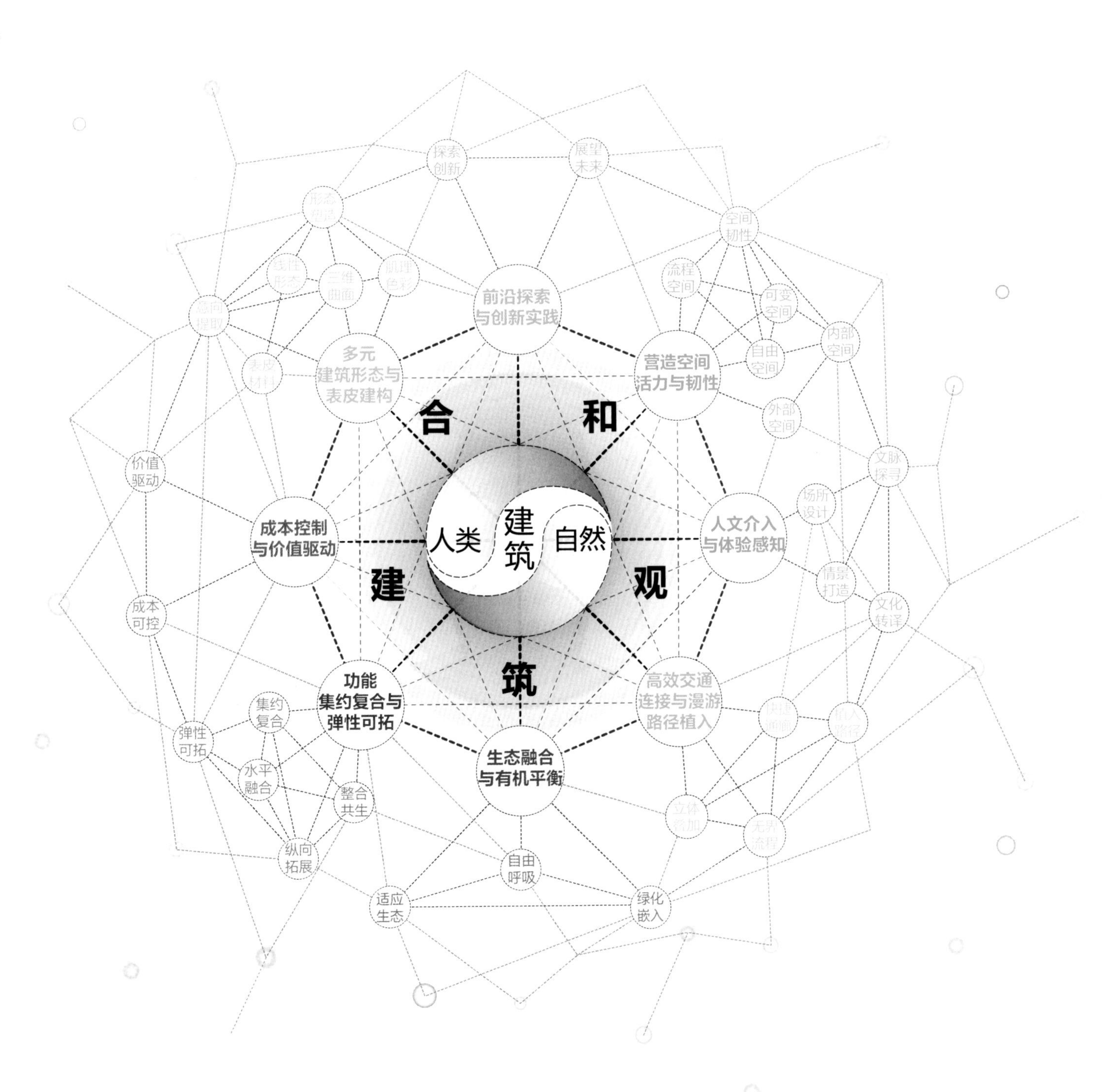

合和建筑观理论体系框架
The theoretical framework of integrated and harmonious architecture

1.1 BACKGROUND

Along with the development of reform and opening up, urban construction in China has witnessed the change from small towns to megacities. In this process, the five dimensions, i.e. regional characteristics, continuity of the concept "integration and harmony", high-density urbanization, diversified modern needs and sustainable development have served as the background for the formation of the theories of integrated and harmonious architecture.

1.1.1 Regional Adaption of Architecture

Long affected by geographical, historical, cultural and other factors, architecture gradually develops some unique regional characteristics. Architectural design should actively respond to the needs of cities, pursue quality, and constantly integrate new technologies to achieve sustainable development and reinterpret the essence of contemporary Lingnan architecture in practice.

1.1.2 Nourishment by the traditional Chinese concept"integration and harmony"

"Integration and harmony" is a typical concept and one of the basic spirit of the traditional Chinese culture. Its influence can also be found in the traditional Chinese construction. Continuing to this day, this traditional concept has led to the pursuit of contemporary Chinese people for balance and harmony, as reflected in some architectural creations.

1.1.3 High-density Urbanization and High-quality Architectural Creation

As forefront of China's urbanization process, Lingnan region has seen the emergence of a number of high-density urban agglomerations. In such urban contexts, buildings vary a lot in both typology and style with the project practice. This evokes the thinking about the appropriate posture for architecture to appear on such high-density sites.

1.1.4 Diversified Demands and Evolution of Architecture Identities in Modern Times

ID means how things get defined in terms of what they are. Architecture, as an important material carrier of human society, has seen different architectural IDs as affected by the needs of different times Nowadays when traditional architectural ID can no longer meet the needs of the times, more diversified architecture positioning and more dynamic factors and dimensions are required.

1.1.5 Sustainable Development of Architecture

Building industry consumes huge amount of energy and it is one of the crucial problems that need a solution in sustainable development. Reflections on current building activities have shown the potentials for improvement in many aspects, from design concepts, strategies, techniques to construction methods. Since China is a country with large population and limited resources, the negative impact of architecture practice is more likely to cause a scale effect.

Architecture practice and theory have been developing simultaneously with the process of urbanization, and the evaluation criteria of architecture are diversifying. This puts forward requirements on adaptability of modern architectural practice, and calls for exploration into a new architecture theory with sustained vitality.

1.2 INTEGRATED AND HARMONIOUS ARCHITECTURE

1.2.1 The Nature of Architecture and the Development of Architectural Factors

In order to develop a theory that is compatible with architectural practice, it is necessary to think about the nature of architecture and the development of its factors.

1)The Nature of Architecture: Human Refuge in Nature

Architecture originated from the basic needs of human beings to shelter from wind and rain. When homo sapiens began to leave mountain caves and trees and live on the ground, the first imperative was to find safe shelters against harsh weather conditions and the threat of wild animals. It was primitive architecture of multiple functions, including living, storage and breeding. It also had easy access and was equipped with some basic architectural factors. Although with the development of human society, the connotation of architecture has been changing under the influence of social values, its essence has always been dealing with the relationship between man and nature. Therefore, architecture exists as a material carrier to deal with such relationship and serves as the link of them.

2)The Development of Architectural Factors

With the development of human civilization, architectural identities have gained more profound connotations gradually, and different civilizations have developed unique architectural connotations. It means architectural views are different in different regions. Today, when architecture has moved towards the interwoven development of globalization and regional characteristics. Apart from traditional architectural factors, such as function, circulation, space and form, it is a consensus that ecological, humanistic and economic factors are also important in architecture based on concerns for environmental protection, humanism and sustainable development.

1.2.2 The Theoretical Core and System of Integrated and Harmonious Architecture

1) The Theoretical Core: The Pursuit of a Balanced State of Integration and Harmony

The core concepts of the Integrated and Harmonious Architecture are integration and harmony. Integration means to integrate, converge and coordinate the extensive architectural factors; specifically to integrate core factors including function, circulation, space and form and modern factors including green building, humanistic considerations, value and innovation and to integrate them into a whole. Harmony, on the other hand, means an orderly and pleasing balance, which is to further create a coordinated organism on the basis of the coordinated architectural factors. The essential concepts draw on the traditional Chinese philosophy of unity of man and nature, the inclusive and innovative spirit of Lingnan architecture and inspirations from the open and diverse context of contemporary international architectural trends, so as, to achieve harmony between the internal and external factors of the architecture and a systematically balanced and stable state of the overall architectural relationships. The Integrated and Harmonious Architecture pursues balance and stability of the internal and external factors of architecture, and views these factors as both relatively independent and mutually influential and interrelated in shaping a dynamically developing theoretical network and, centering around the fundamental issues of human being, architecture and nature, which are the essence of architecture, constituting the theoretical core of the Integrated and Harmonious Architecture.

2) Design Principles: The Composition of Eight-dimensional Factor System

Based on the theoretical core of the Integrated and Harmonious Architecture, a system of eight design principles is proposed for the internal and external architectural factors, namely, intensive, composite, elastic and expandable function; efficient traffic connection and implantation of roaming path; dynamic and resilient space; diversified building form and skin construction; ecological integration and organic balance; humanistic intervention and experiential perception; cost control and value-driven orientation; as well as frontier exploration and innovation practice. These design principles are never unalterable; instead, they are inclusive and open, and like architectural factors, will be constantly evolving along with the times.

3)Design Strategies: Whole Process Design

Design strategies are specific measures further proposed based on design principles and represent a summary of my 40-year experience in architectural design practice. In the functional dimension, strategies such as intensive and composite function, functional integration and symbiosis, horizontal integration, vertical expansion, reservation of planning flexibility, and functional transition and reorganization are proposed to create a building complex that is composite to meet the characteristics of urban development; in traffic dimension, strategies such as multi-level superposition, corridor design, circulation design, borderless process, urban path creation, and building path connection are proposed to create a convenient and smooth traffic system for a building; in spatial dimension, strategies such as space organization, public space implantation, multifunctional space and variable space are proposed to create free and shared spaces with potentials for future development; in the dimension of architectural forms, strategies such as coexistence of diversified building forms, skin construction and intent expression of architectural form are proposed to create building forms with the characteristics of the times that adapt to local context; in the ecological dimension, strategies such as breathable construction, implantation of multi-level greening, and organic adaption to ecology are proposed to create a green building system that is well integrated into the environment, local microclimate, and makes good use of passive green building technologies; in humanistic dimension, strategies such as contextual element, interaction space and scenario perception are proposed to create places with humanistic characteristics and humanized experience; in economic dimension, strategies such as full-life-cycle cost control are proposed to realize social, economic, artistic and scientific values and to create an efficient and sustainable architectural design, operation and maintenance system; in innovation dimension, strategies such as exploration and innovation practice are proposed and future development modes are discussed to explore a new path of architecture.

4)Realization Path: Tools for High-quality Architectural Practice

The realization path is a set of practical tools of Integrated and Harmonious Architecture and a way to meet the needs of the industry in modern times and to ensure the quality of projects. It consists of three aspects: innovative design path under contemporary complex conditions, architectural design data analysis system for complex large-span buildings, and mutual-fed, closed-loop, life-cycle and total-factor design process. The design path aims to propose an optimized path in consideration of project characteristics, advanced concepts of the industry and cutting-edge technologies, which consists of multiple aspects such as interdisciplinary integration, multi-disciplinary and multi-dimensional green design, life-cycle and multi-disciplinary BIM coordination, etc.. The analysis system aims to introduce rational data analysis and building simulation into the design process to ensure realization of the preset goals of buildings, which consists of quantitative building space analysis, multi-dimensional building environment analysis, simulation-based process analysis, etc.. The design process focuses on the whole process of building from design to operation, ensuring each link is controllable and covering planning and design, construction coordination, as well as operation, maintenance and post-evaluation. Based on the experience summary of contemporary architectural practice, the realization path well adapts to the development of architectural industry. It is not only an important technical requirement of architectural practice, but also an inherent guarantee for continuous improvement of building quality.

1.2.3 The Theoretical Characteristics of Integrated and Harmonious Architecture

Three characteristics are desired for the theoretical system of Integrated and Harmonious Architecture. Firstly, it is both theoretical and operation-oriented. The theory should propose not only the concepts of core value and principles, but also the architectural design methodology and design path for guiding the design practice. Secondly, it should convey a holistic view. The theory carries on traditional Chinese philosophy, evolves from the concepts of Lingnan architecture, and draws on inspiration from contemporary Western architectural theories. It deals with multiple dimensions of architecture, from macro aspects to micro aspects and aims to create an integrated and harmonious unity of factors.Third, it should seek for dynamic balance with the change of the times. The theory should be open and expandable, be able to adapt to the sustained development of architectural design, and seek for balance of integration and harmony in dynamic development.

In my nearly four decades of professional practice, I've been mainly engaged in the design of transportation, sports and expo buildings while trying some other architectural typologies as well. These projects I worked on are mostly large public buildings with complex functions and prominent public nature. They serve a large crowd of people and constitute an important part of the urban image, and therefore, the design should be balanced in a multi-dimensional, integrated and comprehensive manner to meet the complex and diverse needs. In recent years, I further explored this harmonious and balanced design on some small-scale buildings. To conclude, it is based on the analysis of contemporary conditions, the study and development of classical architectural concepts, and reflection on architectural practice that I develop and propose the theory of Integrated and Harmonious Architecture. With a framework comprising theoretical core, principles, strategies and paths, and proven performance in professional practices, the theory tries to express understanding of the nature of architecture and the dynamic development of its factors, and to respond to the core issues in architectural design. It is hoped that the theory will contribute to the already abundant architectural theories, and undergo continuous development and refinement in future practice to exert sustained vitality.

CHAPTER II

合和建筑观的多元思考与多维策略

INTEGRATED AND HARMONIOUS ARCHITECTURE: PLURALISTIC REFLECTION AND MULTIDIMENSIONAL STRATEGY

2.1 功能集约复合与弹性可拓

随着社会生活丰富性的增强，建筑功能更趋复杂，其组合及联系由单向简单转变为多向复杂。如何进行复杂功能之间的有效集约组合，在建筑实践中需持续关注及探索，而建筑随地方政策及社会经济的发展，功能需求也在不断变化。在建筑设计实践中，应考虑建筑在不同时期适应功能的动态变化，体现出对高密度城市命题的回应。

2.1.1 集约型的复合功能

建筑功能的集约复合是社会需求多样发展的体现。当代人们对物质与精神两方面的追求使建筑功能也愈发多元，简·雅各布斯在《美国大城市的死与生》一书中提及，城市中心区的建筑主要功能必须多于一个，从而保证不同时间吸引不同的人，提高区域的多样性，激发城市的活力[4]，这种复合性促进了集约型的建筑组团相继出现。

城市化的高速发展和巨量的建设不断挤压城市空间，公共土地资源变得越来越稀缺，在此背景下，为实现当代公共建筑的复杂功能与较大规模建设，满足人口增加后的使用需求，有必要探索集约型的复合建筑体，通过对不同功能的建筑单体进行有效联系，形成整体集约、相互关联的建筑群组。这种建筑群组与城市具有功能同构的特点，能较好回应城市发展诉求，成为城市友好建筑，有效提升土地价值。

1）集约复合

集约复合侧重于建筑单体功能的结合，是节约城市土地、实现资源充分利用的有效方法，通过串联的平台以及连续的屋顶连接不同建筑，使之功能相互结合，形成一种较为集约的功能布局。体育中心

肇庆新区体育中心集约复合的功能布局
Intensive, composite functional layout of Zhaoqing New Area Sports Center, Zhaoqing

往往存在不同功能的场馆，各场馆占地面积大，传统的分散式布局较为浪费土地，若对场馆进行连接，形成一种集约的体育建筑群组，那便可以大大减少对土地的占用。以肇庆新区体育中心为例，该项目包含专业足球场、体育馆主馆和训练馆等三栋主要建筑物，功能齐全，但项目用地面积较为紧张，难以实现分散式布局。方案采用了集群式的设计手法，将各场馆包裹在统一屋面之下，连成一个整体，形成集约的建筑群组，各场馆之间的场地承担起灵活的服务功能，有效节省用地资源。

2）整合共生

整合共生侧重于具有一定互补性的功能体之间的统一布局，通过功能主体之间的合理搭配，形成一种相互支撑的共生关系。这种功能模式常出现于综合体建筑中，建筑的多样功能相互叠加，彼此之间具有一定的联系，合理的功能组合方式是这种建筑类型得以实现高效的重点，需要进行甄别和判断。广州南站核心区 TOD 项目以车站为中心，要求一种整体开发，提供集交通、工作、娱乐和生活于一体的枢纽设计，作为未来TOD枢纽的开发先锋，对功能的整体性要求高，在其实践中，探讨了这种整合的手法，利用商业的连续性特点，以商业功能组织区域其他功能的联系，提升了枢纽片区的整体功能效率。

2.1.2 具有“微城市”场域特征的建筑

如今越来越多的公共建筑逐渐具备了微城市的场域特征。斯坦·艾伦在其著作《点＋线：关于城市的图解和设计》中提出场域状态的概念，从中观的城市视野来看待建筑，将城市的部分功能特征带入

广州南站核心区 TOD 开发与高铁站紧密结合
Close integration of TOD and HSR station in Guangzhou South Railway Station Core Area

了建筑内，描绘出一种类似于城市的建筑状态[5]。这时的建筑就像一座内向型的微缩城市，既拥有了丰富多样的内部功能，又灵活应对了建筑之外的城市环境。

当代大型公共建筑融合了多样复杂的功能和完善的内部设施，除了本身的功能外，还具有了城市体验的功能。具有微城市场域特征的公共建筑表现为内部功能在水平和垂直方向的融合拓展，形成功能种类多样、具有丰富活力的建筑场域，促使人们在建筑当中拥有如在城市中一般的丰富体验。

1）水平融合

大型公共建筑具有一种城市功能内化的趋势，将城市中部分体验集中融合于公共建筑中，可使传统功能较单一的建筑逐渐转变为新的城市活力区域。主要可融合的功能包括商业、文化等城市活力功能类型和社区、出行服务等基础服务功能类型，可围绕建筑中的通行流程来进行布局，使公共建筑内部呈现出独特的功能模式，犹如城市中的街区。机场航站楼，特别是大型枢纽机场航站楼具有高度集中的功能，其最重要且最典型的特征是功能布局需要紧密围绕旅客流程展开，具有组织城市功能的潜力。在广州白云机场 T2 航站楼中，项目从多方面创造了不同于国内其他枢纽机场的商业功能布局模式，创新性地提出了非典型商业综合体概念，以交通作为主线，与旅客主要流程高度整合的集中商业区，使旅客在出行途中可以享受便利的购物环境，并有效增加非航收入，体现建筑的微城市场域特征。

2）纵向拓展

随着城市土地越来越紧张，建筑向空中不断生长，高层、超高层建

广州南站核心区 TOD 集多功能于一体
Integration of various functions in TOD of Guangzhou South Railway Station Core Area

广州白云机场 T2 航站楼非典型商业综合体与流程高度融合
High integration of non-typical commercial complex and process flow in T2 of Guangzhou Baiyun International Airport

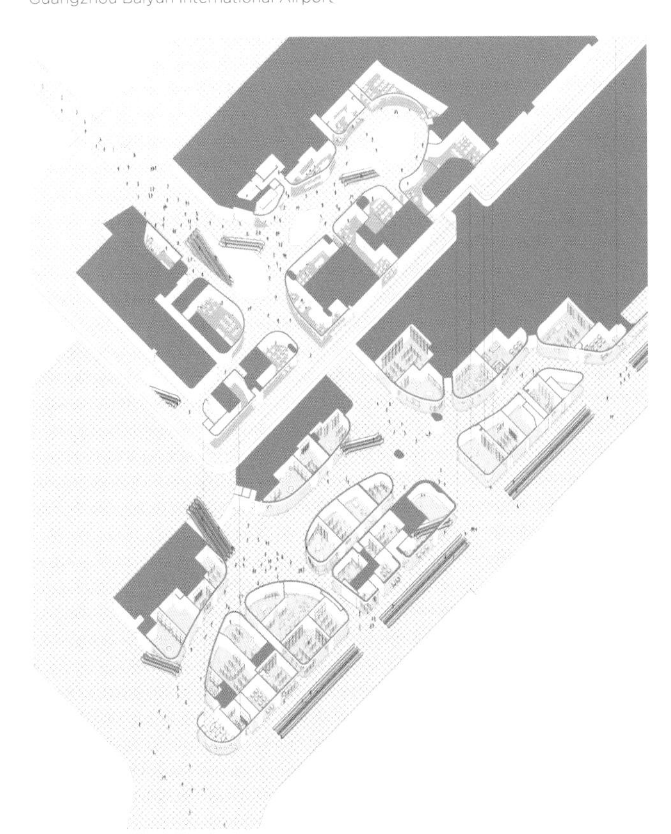

广州白云机场 T2 航站楼以交通为主线的非典型商业综合体示意图
Schematic diagram of traffic-oriented non-typical commercial complex in T2 of Baiyun International Airport, Guangzhou

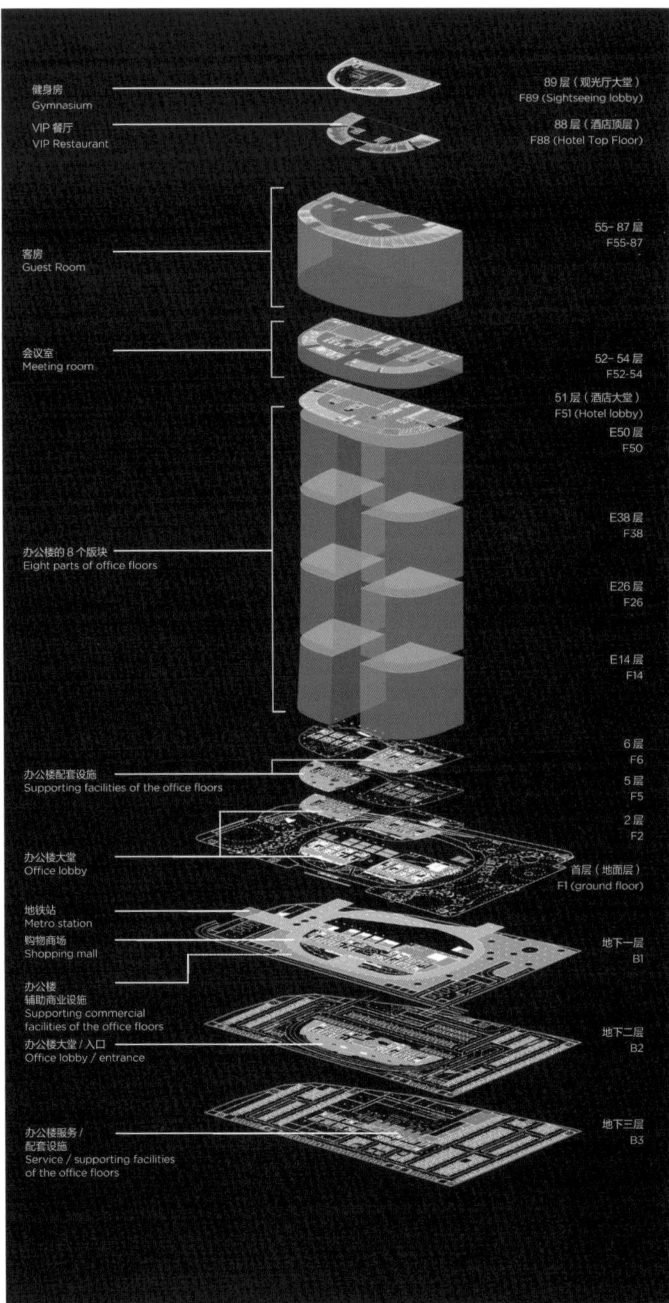

广州西塔投标方案垂直功能分布
Vertical functional layout of the bid proposal for the West Tower, Guangzhou

筑的出现，带动了功能的纵向自由拓展，形成一种新的组织模式。相较于平面布局的建筑，纵向拓展的建筑打破了竖向边界，产生明显的垂直功能联系，可以将城市功能向空中延展，形成立体复杂的功能关系和不同的建筑体验。在高层与超高层项目的实践中，利用建筑体量的组织特点，在建筑中构筑具有特色的功能体块，创造具有开放性、体验性的公共场域，形成竖向的“微城市”。与日本著名建筑师原广司合作的广州西塔投标方案即采用了纵向拓展功能的思路，将建筑分割为两栋办公楼，并在顶部划定一块特色功能区域，结合业态打造出一个集办公、酒店及观光厅一体、具有高度开放性的“天空之城”。分割的两栋办公楼与酒店明确分开，办公楼可以利用酒店的会议室、体育俱乐部等功能，两者既独立又紧密联系，建筑底层设置酒店的入口引导前厅，结合下沉式花园、配套商业街，布置餐饮、咖啡厅、专卖店、旅行社等功能，顶部的观光厅则包括体育俱乐部、周游性餐饮和眺望台，竖向功能联系带来了具有生长特性的立体场域“微城市”。

2.1.3 弹性发展的功能可拓性

公共建筑的发展与地方政策及社会经济发展水平密切相关，因为这种动态关系，在不同时期需适应功能需求的变化。斯图尔特·布兰德在其专著《建筑养成记》中指出：建筑是动态的事物而不是静态的，建筑随时都在发生改变，根据功能需求的变化而调整[6]。强调建筑设计的弹性思维，将时间因素纳入其中，把对于建筑设计的理解扩展到建筑物建成、使用直至未来改造、拓展的全生命周期。

公共建筑作为一种物质生命体，承载了大量社会资源，建筑功能的弹性可拓能有效减少资源消耗，满足可持续发展的需要。规划设计阶段通过对方案进行弹性预留，可以减少分期建设的成本；扩建阶段通过对功能进行灵活的承接重组，可以赋予建筑持续的生命力。

揭阳潮汕机场航站楼可持续发展构型
Sustainable configuration of Jieyang Chaoshan International Airport

1）规划弹性预留

在规划设计阶段对项目的分期建设进行预先安排，并预留相关的功能拓展条件，以满足未来建筑使用容量的发展。以机场航站楼与TOD为代表的一些大型项目具有多期开发的需求，需进行弹性预留。航站楼通常需要随着客流量的变化进行阶段性扩建，以保障不同时期机场的高效运行；TOD开发也常面临轨道与上盖开发的适配，上盖开发基于建筑范围内不同轨道站点的发展规划，进行功能同步预留。揭阳潮汕机场作为干线机场，需要通过更加集约的功能与集中的管理以提高整体效益，航站楼分为一、二、三期建设，一期通过合理规划，在布局、功能流程与设施设备等方面预留弹性可拓的条件，将航站楼二、三期构型进行了多方案设计比选，最终选择了综合最优的“单主楼＋三指廊”构型，为机场近远期的发展提供了规划上的预留，未来扩建后仍然是一个整体的航站楼，极大方便旅客使用和机场运营管理；在深圳机场东枢纽投标方案中，在多条轨道上方设计与枢纽高度融合的TOD开发，对地下多条轨道交通对比分析，提出地铁、城际铁路、国铁同基坑同期建设的方案，建造成本低，同时满足未来上盖物业不同时期的改造与扩建。

2）功能承接重组

在建筑扩建时，对新旧部分的功能进行重新梳理，根据扩建后的使用需要进行功能的再整合，建立新的秩序。在具体操作中应根据功能布局实际情况灵活调整，确保调整前后功能合理性。揭阳潮汕机场航站楼的二期扩建基于一期的统一规划，结合发展现状对航站楼局部构型进行拓展，南边扩建国内航站楼，北边扩建国际指廊，通过对现状航站楼的国内流程改造，将现状功能房间及设施设备进行“拆分”与“重组”，由分流改为混流，共用服务设施等功能资源，节省管理成本，实现改扩建后航站楼整体效率的进一步提升。

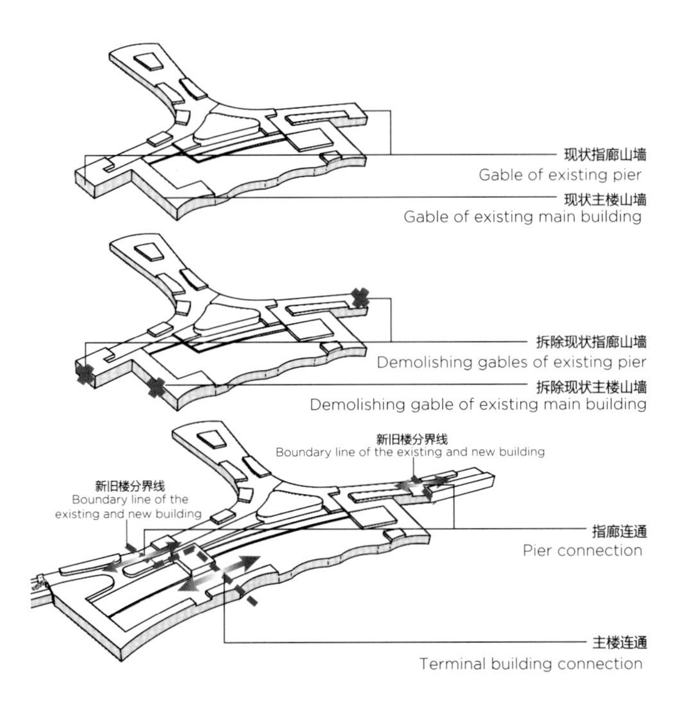

新旧主楼及指廊衔接
Pier connections of the new and existing main buildings

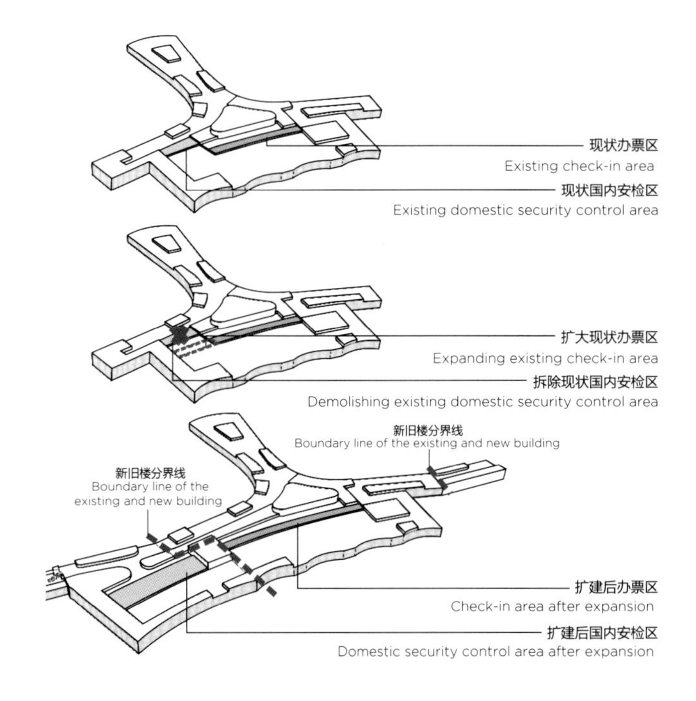

设施设备拆分与重组
Division and reorganization of facilities and MEP system

2.1 INTENSIVE, COMPOSITE, ELASTIC AND EXPANDABLE FUNCTIONS

Buildings functions tend to be more complicated due to the diversified societal life. How to effectively and intensively combine complex functions is yet to be identified through constant concern and exploration in architectural practice. Therefore, the adaption of buildings to dynamic changes of functions in different periods should be considered in architectural design practice to respond to the proposition of high-density cities.

2.1.1 Intensive and Composite Functions

People's physical and spiritual quests in contemporary society work in tandem to diversify architectural functions. Jane Jacobs mentioned in her book *The Death and Life of Great American Cities that* there must be more than one main building function in an urban center, so that different people can be attracted at different times. Such composite functions fueled the successive emergence of intensive building clusters.

As rapid urbanization and massive construction continue to squeeze urban spaces and public land resources become increasingly scarce, it is necessary to explore buildings of intensive and composite functions, and form overall intensive and interrelated building clusters by effectively connecting individual buildings of different functions. This kind of building clusters are marked by functions that can be shared by the city. They can better respond to the demands of urban development, create city-friendly buildings, and effectively enhance land value.

1) Intensive Composition

Intensive integration focuses on the functional integration of individual buildings. As an effective way of saving urban land and exploiting resources, it combines the functions of different buildings via interconnected platforms and continuous roofs. Sports centers usually contain different functional venues, which makes the conventional decentralized layout a waste of land. With proper connections, the venues may form an intensive sports building cluster and largely reduce land occupancy. Zhaoqing New Area Sports Center is a good example. Taking the cluster design approach, the three main buildings of the fully-functional project are brought together under one same roof to form an intensive cluster, which effectively saves land resources to adapt to the tight site area.

2) Integration and Symbiosis

Efficient composition emphasizes the unified layout of complementary functional blocks. It can improve the service capability and quality of functional blocks and rationally combine them to develop a symbiotic relationship of mutual support. This functional model is often seen in complex buildings, where multiple functions overlap each other and present principal-subordinate as well as service relations to a certain extent. A reasonable functional combination is the key to high efficiency of this building typology. The design of the TOD project in Guangzhou South Railway Station Core Area probes into this highly efficient and mixed approach. As a station-centered hub integrating transportation, work, entertainment and life, the TOD project is highly demanding for functional composition. The overall functional efficiency of the hub area is improved by linking up all functions via commercial function that is continuous in nature.

2.1.2 Buildings with Mini City Field Characteristics

Nowadays, an increasing number of public buildings have developed mini city field characteristics. In his book Points and Lines: Diagrams and Projects for the City, Stan Allen puts forward the concept of field conditions, introducing some features of urban functions and spaces into a building, making it a miniature city that has rich interior functions while flexibly responding to the dynamic, diverse urban environment around it.

Contemporary large public buildings integrate varied, complex functions and perfect interior facilities. Public buildings with mini city field characteristics demonstrating the horizontal and vertical integration and expansion of interior functions can foster a type of building field marked by diverse functions and great vitality, engaging people with rich experiences as in a city.

1) Horizontal Integration

Large public buildings now tend to internalize urban functions. By centralizing some urban experiences inside, it's possible to turn traditional public buildings with onefold function into new urban vitality areas. Different types of urban vitality functions and basic service functions can be placed around the circulation routes inside public buildings to resemble urban blocks. Airport terminals where functions are highly centralized around passenger handling processes have the potential for accommodating urban functions. The design of T2, Guangzhou Baiyun International Airport features a commercial layout pattern different in many aspects from that of other domestic hub airports, as it innovatively creates a centralized commercial area highly integrated with the main passenger handling processes. By doing so, a convenient shopping environment is provided to passengers, which effectively increases the non-aviation revenue of the airport and fully reflects the mini city field characteristics of public buildings.

2) Vertical Expansion

As urban land for construction gradually decreases, buildings start to grow toward the sky. Vertically expanded buildings break the vertical boundary and extend urban functions into the air, adding a new attribute to public buildings. In the practice of high-rise and super high-rise projects, the architect can follow the characteristics of building volume organization to create open, experiential public fields and vertical mini cities inside buildings. The bid proposal for Guangzhou West Tower, aka Guangzhou International Finance Centre, a project jointly designed by GDAD and the renowned Japanese architect Hiroshi Hara is an example of vertical expansion of building functions. The tower is divided into two office buildings and a distinctive functional area is planned on the top, i.e. a highly open "sky city" integrating office, hotel and observation hall in combination with business formats. The two office buildings are clearly separated from the hotel but closely connected with it in terms of vertical function, making for a growing mini city featured with vertical field.

2.1.3 Functional Expansion of Elastic Development

Public buildings need to adapt to changing functional requirements in different periods. Stewart Brand points out in How Buildings Learn that "buildings are dynamic instead of static." He emphasizes elastic thinking in architectural design and incorporates the factor of time, extending the understanding of architectural design to the whole life cycle of buildings from completion to use, future renovation and expansion.

As a material organism, public buildings carry plentiful social resources. The elastic expansion of building functions can effectively reduce resource consumption and enable sustainable development. Elastic reservation of design conditions in the planning stage can slash the cost of phased construction, while flexible transition and recombination of functions in the expansion stage can inject lasting vitality into the building.

1) Reservation of Planning Elasticity

Reserving planning elasticity means to define development phases early in the planning stage, and reserve conditions for functional expansion as building capacity grows. Some large-scale projects, such as airport terminals and TODs, have the need for multi-phase development and the reservation of elasticity. Airport terminals usually need to be expanded by phases as passenger traffic grows, while TODs are often faced with the adaption of rail track to superstructure development. Our planning for Jieyang Chaoshan International Airport Terminal in Phase I offers planning guidance for the near and long-term development of the airport by reserving elastic expansion conditions and employing the most advantageous configuration pattern. Our bid proposal for Shenzhen Airport East Integrated transportation Hub, with a TOD project above and highly integrated with multiple rail tracks, puts forward the synchronous construction of metro line, intercity railway and national railway together with the foundation pit, which cuts the construction cost and meets the requirements of superstructure reconstruction and expansion in different periods in the future.

2) Function Transition and Recombination

Function transition and recombination means to review the new and existing functions of a building to be expanded and re-integrate them for a new order according to the post-expansion uses. The Phase II expansion of Jieyang Chaoshan International Airport Terminal is based on the unified planning of Phase II. Partial configuration of the terminal is expanded in combination with the development status quo, which includes the southward expansion of the domestic terminal and the northward expansion of the international pier. Through the transformation of domestic process in the terminal, the existing functional rooms, facilities and equipment are "split" and "recombined", the way of passenger handling is changed from separate flows to mixed flow, and the service facilities are made shareable, which combine to save management costs.

2.2 高效交通连接与漫游路径植入

公共交通随着城市化进程同步发展。如今城市公交体系日益庞杂，换乘方式日益多样。随着交通需求增多，交通目的地越来越分散，这无疑对交通整合和出行体验提出了更高的需求；此外，建筑的交通流线愈加复杂，室内外通行流程丰富多样，路径体验的人性化需求也在逐步增加。通过重新打造交通系统，实现建筑内外立体交汇换乘、无界感知通行与漫游化交通体验，进行“站、人、城”多维度的融合，提升联系效率，丰富通行体验。

2.2.1 立体交汇集散

工业革命推动了大都市的发展，大量增加的人口和二维扩张的城市导致地面交通压力陡增。为解决此问题，《雅典宪章》从适应现代交通工具发展的角度，提出了交通高效发展的概念与立体城市的考量，城市要素交汇与叠加，拉开了城市立体交通的序幕 [7]。

随着城市中不同交通类型的出现，人们对交通间的高效集散提出了更高要求，交通与城市之间的结合方式不断发展。纵观国内外大型交通枢纽，这种改变已在进行——以日本涩谷为例，其作为东京最繁忙的街区之一，经历了 20 多年来城市和交通交汇成长的过程，铁道、地铁、公交、步行系统与城市商业在水平和垂直方向融合互嵌，形成了日流量超过 300 万人次的高效交通网络。如今，以“多式联运”为代表的发展模式已成为时代趋势，随着大湾区枢纽实践进程的推进，高效联运的交通枢纽不断涌现。交通枢纽容纳的交通方式不断增加，枢纽规模不断扩大，不仅注重枢纽内部各类交通方式之间的高效疏解与换乘，也注重枢纽与周边片区之间高效的步行联系。

广州白云机场 T1 航站楼主楼办票大厅
Check-in Hall of T1 of Guangzhou Baiyun International Airport

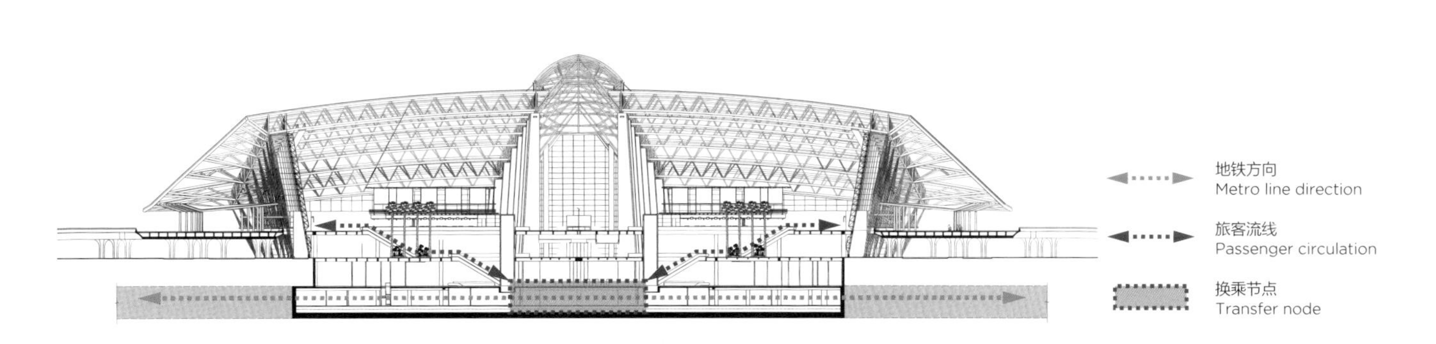

广州白云机场 T1 航站楼与地铁垂直叠加
T1 of Guangzhou Baiyun International Airport vertically stacked on top of the metro line

广州白云机场 T2 航站楼主楼与交通中心“面对面”联系
Face-to-face connection between main building and GTC of T2, Guangzhou Baiyun International Airport

广州白云机场 T3 航站楼主楼与交通中心以换乘大厅衔接
T3 connected with GTC via transfer hall in Guangzhou Baiyun International Airport

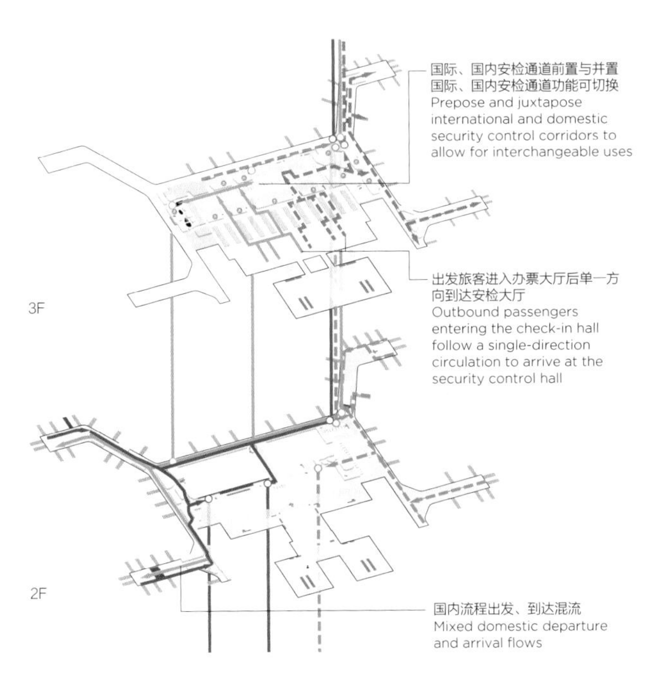

广州白云机场 T2 航站楼便捷的流程设计
The convenient process design of T2 termind of Guangzhou Baiyun Airport

1）立体叠加快速转换
传统地面交通方式难以满足当代交通多样化发展的需求，特别是对于交通枢纽建筑，往往需要在有限的土地容纳多种类型的交通，并满足不同交通之间的换乘。通过不同交通的立体叠加，设置垂直交通核、换乘通廊或换乘大厅进行人流疏解，可充分利用土地资源，并缩短旅客换乘距离，实现快速转换。在广州白云机场项目中，T1 航站楼主楼与轨道交通垂直叠加，形成高效的换乘关系，主楼地下二层引入地铁，出港旅客乘地铁到达后进入主楼中庭，搭乘电扶梯直达三层办票大厅；T2 航站楼与 T3 航站楼在楼前设置交通中心综合体，集地下轨道交通、道路交通多种交通方式于一体，并设置旅客换乘通道与航站楼衔接；在深圳机场东枢纽投标方案中，将高铁、城际、地铁、APM 和各类道路交通方式分类梳理，通过地下、地面、高架等不同形式进行立体叠加，并在众多轨道交通交汇的站台层上方（负二层）设置了综合换乘大厅，不同交通方式的旅客均能方便通过电扶梯到达换乘大厅，进而选择不同的交通换乘方式。

2）快捷通廊接驳片区
在高密度城市空间背景下，交通枢纽与城市的融合度越来越高，表现出站城一体的特征。多样化的交通方式聚集后，对片区的地面交通疏解提出了更高的要求。通过在枢纽中设置专用步行通道接驳周边片区，鼓励片区人群步行来往枢纽，能有效减少交通压力。专用步行通道实现了人车流线的分离，确保效率与安全。在深圳机场东枢纽投标方案中，航空、轨道、地面交通流线复杂，需要清晰直接的专用步行路径连通片区与枢纽。结合枢纽交通组织，在负一层、负二层换乘大厅沿多个方向设置与周边片区连通的城市通廊，满足周边居民、办公等多类人员快速往返枢纽内部。地下通道也为人们提供了遮阳避雨的条件，与岭南地区气候条件十分契合。

2.2.2 无界感知流程
一体化的交通流程是从长远的角度去思考枢纽站的未来发展方向。对于交通枢纽，在基本交通格局的基础上，需要对通行流程进行一体化设计理念，实现未来枢纽站的通行界面融合。20 世纪 80 年代以来，众多国际大都市陆续运用多种交通一体化的新模式，以弱化流程的界面感知，解决复杂交通通行流程不便的问题，诞生了诸多成功的典范，也成为我国当下交通流程建设的发展趋势。

在站城“一体化开发”的契机下，交通的无界通行实践成为可能，为进一步发挥公共交通的效率优势，要保证交通系统的整体性与便利性。一个大型的复合枢纽机场航站楼拥有各种类型的旅客流程，多种交通流程错综复杂，或交汇、或联系，在对交通流程设计中，应设置形成连续性的通行体验，使交通内部通行融合，实现“无界”的体验。未来一段时间，车站将与交通技术更加高度结合，流程设计更加紧密无缝衔接，简洁高效的无感流程设计将成为路径畅达的重点，实现交通无界通行。

1）高效流程设计消解交通边界
交通建筑的流程设计是重中之重，高效的流程设计可减少旅客滞留，消解交通边界。通过综合策略搭建合理的内部流程框架，优化流程，提高引导清晰度，降低识别难度，实现流程明确、效率为先的目标。广州白云机场 T2 航站楼的交通流程可作典型例证，T2 航站楼采用单一方向的旅客流程策略，让出发旅客进入办票大厅后只有向前一个方向，到达旅客提取行李后通过统一的到达厅与交通中心连通，国内流程采用混流设计，到达出发旅客共用了商业、卫生间、问询柜台等服务设施，提高了资源利用率；通过国内 / 国际可切换的混合机位流程设置，可同时分层候机，切换登机，提高了机位使用率；通过国内混流候机层、国际分流到达层在二层同层设置，国际转国内可全程平层解决，方便快捷；中转共用始发、终到流程的检查场地资源，不但节约场地和人力资源，还可降低标识引导的复杂性和减少旅客选择点。

2）智能智慧技术打造无界感知

在交通建筑中，智能智慧技术的融合能有效提升旅客的通行体验，实现无感化通关。自助托运使旅客轻松出行，自助关卡能减少旅客通关停留时间，从而改善流程衔接，打破流程之间的边界，带来无界通行感知。在广州白云机场 T2 航站楼中，提供全流程自助出行服务，采用多种旅客智能设备。围绕自助值机、自助行李托运、自助登机、自助通关、“无纸化”出行等开展全方位建设，实现无感化通关，为实现全面自助的智慧机场发展创造了条件。

2.2.3 植入漫游路径

20 世纪初期，交通枢纽的漫游路径设计已在国际上展开探索，从世界上第一个枢纽百货——阪急百货店的出现，到具有艺术活力的纽约中央车站，见证了国际上探索交通枢纽城市漫游的历程。交通枢纽具有多元业态，在不同的枢纽站和沿线，纷纷建立了住宅、学校、公园、动物园、剧场、酒店、体育场等大众娱乐和生活设施，业态丰富，极具城市活力。

当代交通枢纽具有与城市更加融合的趋势，强化城市功能，弱化交通特征，交通枢纽不仅提供功能化通行，还更多地满足了人们日常交往需求，成为具有漫游特征的城市活力场所。具有未来气质的超级枢纽正在加速发展，“漫游式枢纽”等新概念出现，表明枢纽内

广州白云机场 T2 航站楼办票大厅中自助值机与自助行李托运
Self-service check-in and baggage check-in of T2 Guangzhou Baiyun International Airport

部正继续经历“去枢纽功能化”的过程，其内部明确的功能条块与边界也处于消融的过程中。

1）城市路径建构

大型交通枢纽对城市风貌的影响很大，其往往是城市自然、历史、人文、经济交流交汇之处，也是门户名片和未来发展的核心引擎，因其体量往往较大，处理不佳容易形成城市联系的割裂。通过以慢行为主的泛目的性路径设计，可使交通枢纽成为联系城市的漫游场所，结合服务业态和门户客厅的定位，承担重要的城市职责。将枢纽缝合进周边城市片区的统一发展中，成为城市漫游路径的一部分，完善了城市整体漫游体系。以深圳机场东枢纽投标方案为例，其紧邻珠江口，面向伶仃洋，服务大湾区，通过重塑枢纽两侧和周边的

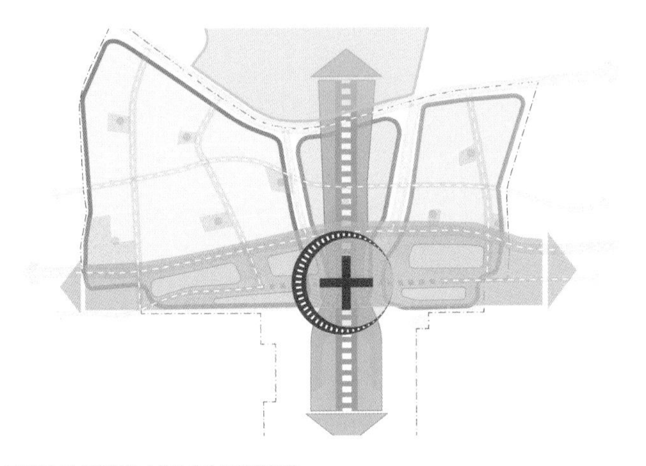

深圳机场东枢纽投标方案十字主轴漫游路径
Cross-shaped main axis roaming paths in bid proposal for Shenzhen Airport East Integrated Transport Hub

深圳机场东枢纽投标方案城市漫游路径建构
Urban roaming path in bid proposal for Shenzhen Airport East Transport Hub

城市结构，确立山海航城新轴线，以一条自然绿带串联起东边凤凰山和西侧海面，让城市漫游步道一直往西蜿蜒至枢纽平台；通过南北轴步行网络串联起枢纽与周边开发，站城产高度融合，提供具有多样城市业态体验的漫游路径；通过建构一个舒适宜人的枢纽环境，最大化向城市开放，将枢纽融于城市，成为城市通行的缝合器和活力源，提供丰富多样的城市漫游路径，带来独特的城市漫游体验。

2）建筑路径串联

公共建筑面向城市进一步开放，一种内外多样漫游的新建筑形式正在形成，漫游路径串联建筑节点和城市，相互联系。在公共建筑内部打造可供体验的漫游路径，通过地面广场、中部休息平台、屋顶花园等区域，形成与建筑外部路径相联系的路径节点，将建筑内外进行连通，补充城市漫游系统，丰富城市漫游体验。以深圳改革开放展览馆投标方案为例，作为片区重要的标志性公共建筑，将漫游序列的设计围绕其展开，形成了内外多级的漫游路径系统，也呼应了香蜜湖山海生态通廊。建筑内部螺旋上升的主要路径轴线与室外直连屋顶的漫游路径轴，连通地面、室内各层与屋顶，形成回环；结合观众浏览路径、改革开放展览路径、生态体验路径、社会教育体验路径等主要路径，层层递进，寓意非凡。方案打破了交通线性边界，协调了建筑与城市的关系，多样慢行路径形成串联城市公共空间与建筑内部的慢行步道，营造出完善的城市漫游系统，通过对枢纽所在片区的业态进行协调，以路径形成脉络，打造出文化、生态、商业相融通的丰富路径体验。

深圳改革开放展览馆投标方案建筑内外串联游径
Interconnected indoor & outdoor roaming paths in bid proposal for Shenzhen Reform and Opening up Exhibition Hall

2.2 EFFICIENT TRAFFIC CONNECTION AND ROAMING PATH IMPLANTATION

Public transportation develops in the same pace with urbanization. Today, the urban public transportation system is getting more and more complex with increasing transfer options. The growing travel demands of people means increasingly decentralized transportation destinations, which undoubtedly requires a higher degree of transportation integration and better travel experience. Moreover, the demand for a people-oriented path experience is also escalating as result of the more complex circulation routes inside buildings and more diverse interior/exterior traffic flows. The reconstruction of the transportation system can help realize multi-level transfer between the interior and exterior of buildings, borderless mobility, roaming-like traffic experience and multi-dimensional integration of "station, people and city", thus improving the communication efficiency and enriching mobility experience.

2.2.1 Multi-level Confluence and Distribution

The industrial revolution propelled the development of metropolises, while the surging population and urban sprawl imposed great pressure to ground transportation. To address this problem, the Athens Charter proposed the concept of efficient transportation development and the consideration of vertical cities to adapt to development of modern transportation means. The confluence and superposition of urban elements heralded the era of multi-level urban transportation.

The emergence of different urban transportation means leads to the confluence and superposition of transportation and city, and brings about efficient communications in modern cities. With high population density, developed economy and perfect match between transportation development and demands, larger cities have the potential to conduct a magnitude of multi-level transportation practices. Nowadays, the development pattern represented by multimodal transportation has become the trend of the times, and efficient multimodal transportation hubs are gradually taking shape.

1) Multi-level Superposition and Rapid Transfer

The traditional ground transportation can hardly accommodate the transportation demands today, while rapid transfer is still possible through the multi-level superposition of different transportation means, and the provision of vertical circulation cores, transfer corridors or transfer halls for passenger distribution. In the design for Guangzhou Baiyun International Airport, T1 building is vertically overlapped with the rail transit to enable efficient transfer; T2 and T3 are both provided with a front GTC to integrate underground rail transit and road transportation, and passenger transfer channels to link with the terminals. In the bid proposal for Shenzhen Airport East Integrated transportation Hub, various means of road transportation are defined and systemized for multi-level superposition; a transfer hall is provided above the platform floor where the rail tracks meet.

2) Connection via Rapid Passage Corridors

In high-density urban spaces, transportation hubs and cities are more integrated. By providing dedicated pedestrian passages between a transportation hub and its surrounding areas and encouraging people to walk to and from the hub, the traffic pressure can be effectively mitigated. In the bid proposal for Shenzhen Airport East Integrated transportation Hub, urban corridors are provided from the transfer halls on B1 and B2 to the surrounding areas based on the traffic organization, allowing residents, office workers and other pedestrians to move rapidly to and from the hub. Underground passages are also provided for sun shading and rain shelter, well responding to the climate in Lingnan region.

2.2.2 Borderless Process

Transportation hubs require a huge and irreversible investment in underground space construction, and an integrated design concept for the transfer spaces. Since the 1980s, many international metropolises have realized integrated transfer between several transportation means to tackle inconvenience caused by complicated transportation. Such practice has been proved feasible by many successful cases and become the current trend of transportation construction and development in China.

The station-city integrated development makes the borderless traffic experience possible. A large airport terminal hub with composite functions involves different types of passenger processes and complicated traffic processes. Zero transfer is possible with various closed-loop transfer routes. In the future, stations will be more closely integrated with the transportation technologies where imperceptible process design will become a focus.

1) Efficient Process Design to Eliminate Transportation Borders

Process design is key to transportation buildings. Efficient process design can minimize passenger retention and eliminate traffic boundaries. A reasonable internal process framework built through comprehensive strategies can optimize processes, ensure clearer signage and recognisability, and eventually realize clear processes and high efficiency. T2, Guangzhou Baiyun International Airport adopts a one-way passenger process strategy, which means departing passengers move toward only one direction after they enter the check-in hall, and arriving passengers arrive at one unified arrival hall leading to the GTC after they pick up their baggage. The domestic mixed-flow waiting area and the international separate-flow arrival area are both placed on F2 for easy transfer between international and domestic flights. The inspection area of the departure and arrival processes is shared by the transfer process, which saves space and human resources, simplifies the guidance signage system and reduces the selection points of passengers.

2) Intelligent Technologies for Borderless Experience

The incorporation of intelligent technologies into transportation buildings can effectively upgrade the traffic experience of passengers and enable contactless security clearance. Self-service baggage check-in allows passengers to travel easyly, while self-service check reduces their stay time at the checkpoints, both ensuring smooth, borderless process and experience. In T2,

Guangzhou Baiyun International Airport, whole-process self-service travel is enabled by various intelligent passenger devices. Comprehensive self-service development in terms of check-in, baggage check-in, boarding, security clearance, paperless travel etc., makes contactless customer clearance possible, laying foundation for creating a smart airport of comprehensive self-service.

2.2.3 Roaming Path Implantation

Architects started probing into the roaming path design of transportation hubs early at the beginning of the 20th century. The urban roaming history of transportation hubs around the world has been witnessed by the first hub department store, i.e., Hankyu Department Store and the artistically vibrant New York Central Station. Public entertainment and living facilities mushrooming at different hubs and along different lines show an abundant mix of business formats and inject great vitality to the cities.

Contemporary transportation hubs tend to be further integrated with cities, generating dynamic urban places with roaming characteristics. The emergence of new concepts such as "roaming hub" indicates that hubs are undergoing "de-hub functionalization" inside.

1) Urban Path Creation

Large transportation hubs often have gigantic building volumes, inappropriate treatment of which is likely to sever urban links, while the design of general-purpose paths mainly for slow traffic can turn a transportation hub into a roaming place to connect with the city. By creating roaming paths in a hub, it's possible to stitch the hub into the unified development of its surrounding urban areas and make it part of urban roaming paths. For example, in the bid proposal for Shenzhen Airport East Integrated Transportation Hub, a natural green belt is planned to connect the Fenghuang Mt. in the east and the sea in the west, extending the urban roaming trails all the way westward to the hub platform. By fostering a pleasant hub environment and opening to the city to the maximum, the hub is turned into a suture and vitality source for urban transportation to bring a unique urban roaming system.

2) Building Paths Connection

By devising experiential roaming paths inside a public building and creating path nodes connected with the exterior paths of the building via ground square, resting platforms on middle floors, roof garden, etc., it's possible to connect the inside and outside of the building to further complete the urban roaming system and enrich the urban roaming experience. The bid proposal for Shenzhen Reform and Opening up Exhibition Hall offers a good example. Inside the building, a main path axis is created to spiral upward and an outdoor roaming path axis is provided to directly connect to the roof. By creating paths respectively for visitors, the reform and opening up exhibition, ecological experience and social education experience, the project offers slow traffic paths linking up urban public spaces and the interior of the building, and establishes a perfect urban roaming system with diverse paths of culture, ecology and commerce.

2.3 营造空间活力与韧性

空间是人们体验建筑的基本途径，也是评价一个建筑作品好坏的重要标准，建筑通过空间的组织，对活动场所进行多维要义的和谐演绎。现代建筑空间发展出更自由的组织、更公共的属性与更具适应性的能力，通过组织连贯而各具特征的空间，为人们提供丰富多样的空间体验，表现出空间的多维要义。

2.3.1 空间动态组织

空间组织的探索早已有之。19 世纪初，法国建筑学家迪朗在对建筑类型进行分析时，曾剥离建筑装饰的细节，探讨了建筑空间组合的模式，其梳理出的空间原型图解至今依然可以发挥作用，这说明建筑的空间组织具有一定程度的稳定性。随着建筑功能与类型的不同，建筑空间的组织方式也存在一定差别。

如今建筑内涵不断发展，建筑功能更加复合，空间组织也出现了更多样的变化，形成了集聚式、线性、单元式、放射状、集簇式等多种丰富的组织形式。基于交通与体育类建筑的功能和流程分析，空间的组织具有两种较为明显的类型，一种是具有逻辑流程的空间组织，另一种是自由灵活的空间组织，这两种空间组织各有特征，流程式空间利于层层递进，自由式空间利于形成自由主动的使用体验。大型公共建筑因功能复杂，服务人群多样，具有多样的空间关系，通过这两种空间组织类型的设置，可打造出具有逻辑与活力的建筑空间。

1）流程式空间组织

流程式空间常体现在具有明显流程顺序的建筑类型中，体现出一定的逻辑性，在这种建筑设计中，应注意体现清晰流程的层进式空间。以交通建筑为例，其空间组织类型明确，机场有明确的办票、安检、候机、出港、到港流程，结合配套服务，在空间设计时需对旅客进行主体连贯、兼顾停留的引导，形成丰富的流程空间节点。广州白云机场 T2 航站楼在建筑空间设计中，针对旅客流程特点，进行具有动感、保持导向性的空间组织。为适应各流程节点特征，分别打造出宽阔通透及具有引导性的出发厅、相对停留的安检厅、具有线性特征的旅客候机区，各区域适合流程特征形成空间的递进组织。

广州白云机场 T2 航站楼具有线性特征的旅客指廊空间
Linear passenger pier area of T2, Guangzhou Baiyun International Airport

2）自由式空间组织

自由式空间常出现在具有丰富活动的建筑类型中，体现出一定的自由度，相对于流程式空间，更具有相对独立性，有利于多样化的流程组织，在这种建筑设计中，应注意体现自由流动的多节点空间打造。这种空间可见于体育馆、演艺中心等建筑类型中，在广州亚运馆的早期实践中，创造性地采取了三维流动空间的设计策略，将四个场馆围绕中部空间进行并联组织，创造连续流动的非均质空间，具有多种使用顺序，形成整体自由、外部成环、局部具有流程的空间组成方式；花都东风体育馆、肇庆新区体育中心也采用了类似的方式，围绕中心形成自由流动的组织。这种做法让建筑空间具有了主动探索的趣味，自由式空间的自由活力体现得淋漓尽致。

广州白云机场 T2 航站楼具有导向性的流程式空间组织
Oriented process-based space organization of T2, Guangzhou Baiyun International Airport

广州亚运馆层次丰富、自由的空间组织
Hierarchical, free-flowing space organization of Guangzhou Asian Games Gymnasium

2.3.2 公共空间置入

公共空间的设计对社会活力的重要性早已不言而喻。杨·盖尔曾提出人类社会活动特别依赖于户外公共空间的设置[8]，随着当代建筑空间的进一步开放，建筑公共空间也出现了多种多样的形式，对其品质的打造，也呈现出新的特征。

与建筑外部公共空间的作用类似，建筑内部的公共空间同样有助于人们的交往活动，这两者在公共建筑中均具有激发活力的意义。外部公共空间体现出一种更全方位的开放，相比之下，内部公共空间体现为一种有限度的开放。在一些公共建筑的设计实践中，可以基于基本功能空间使用，将空间界面适度开放，设置良好的内外公共空间环境，综合提升建筑的活力。

1）外部公共空间

外部公共空间为建筑本体以外，提供人们日常生活与公共活动的空间，具有较高的公共性。利用建筑周边环境设置多种外部公共空间，在空间中划分出公共、私密等不同区域类型，提供人们自由选择活动区域的可能性，促进公共空间的人群聚集，并形成地区活力名片。以德胜体育中心为例，建筑包括主体部分的“三馆一场”、配套商业、停车库和周边景观空间，建筑在有限的用地条件下，设置了室外活动平台、地面广场、二层平台、架空层和景观屋面等多个外部公共空间，形态丰富、疏密有致，公共空间各自具有明显特征，体现出丰富的层次，可供集散、运动、漫步、休憩、展示、宣传、商业等活动使用，具有活力。

2）内部公共空间

建筑内部公共空间在一定程度上同样具有面向公众服务的特征，同时

顺德德胜体育中心多层次的外部公共空间
Hierarchical outdoor public spaces of Shunde Desheng Sports Center

存在着与建筑主要功能的交互关系，具有适度的公共性。根据与建筑的位置关系，有相对独立的室内公共空间和完全融入室内的公共空间，随着人性需求的提升，设置内部公共空间成为显著提升建筑活力的方式。在广州西塔投标方案中，设置了一处具有显著公共性的顶层观光厅，既与建筑内部相互连接，又可直接与城市地面联系，形成相对独立的立体空间，成为一处城市客厅；广州白云机场 T2 航站楼则通过在大厅中创造一处文化广场，通过构件的限定形成与大厅具有一定差别的局部公共空间，激活了空间展示、宣传、交往等多样使用。

2.3.3 灵活空间适配

空间的利用方式不应被完全限定。密斯·凡·德·罗曾提出：“建筑物服务的目的经常会发生变化，但我们不能把建筑物拆掉”，并提出通用空间的理论，以使建筑空间可以满足功能变化的需要；1960 年，新陈代谢派在宣言中提出：“城市和建筑不是静止的，而是像生物新陈代谢那样处于动态发展中”[9]，黑川纪章在 1972 年推出中银舱体大楼，进一步带来了建筑有机更新的思想浪潮。建筑随需求变化的适应性具有发展潜能，已在众多项目中得到实践，体现出其生命力。

时至今日，建筑空间随需求而变的能力已成为建筑空间设计的重要方面，这不仅反映出人们对空间认识的不断深化，同时意味着空间的时代适应性在不断增强，空间利用多样意味着建筑空间一方面呈现出一种多义性，具有本身的多功能使用潜力，另一方面可随时代发展产生一定的可变与适应。随着社会经济的发展，准确预测空间的使用要求几乎不可能，基于这种考量，一方面需要使建筑适应多功能的需要，另一方面可允许建筑空间进行变化和调整，这要建筑

广州白云机场 T2 航站楼办票大厅中嵌入文化广场公共空间
Cultural square public space embedded into check-in hall of T2, Guangzhou Baiyun International Airport

空间具有一种可以在不同时间、根据不同需求满足不同使用的特点，并有利于建筑自身的更新。

1）空间多功能利用

设置多功能空间是提高建筑使用效率和适应性、优化建筑资源利用的有效手段。考虑到建筑可能面临计划外的使用方式，通过设置多功能空间，可保证一定程度的灵活性；除此以外，应将传统的房间分类观念转变为功能分类观念，根据功能的使用特点和使用频率差异，可以由同一个空间承载不同的功能使用。在体育建筑中，针对赛时的专业场地需要和赛后的灵活使用，场馆需要具备多功能的特征，以广州亚运馆为例，场馆内空间可根据需要改造成其他类型场馆，体操馆可灵活改动为篮球馆，并被设计成可独立对外营运的场所，成为群众体育的基地；亚运历史展览馆在赛后保留部分展览功能，其余部分可引入创意商业；室外空间在赛时起到集散的作用，赛后作为市民活动空间，在实际运营中，室外灰空间用于举办多种活动时使用方便，起到了多功能的作用。

2）空间的韧性

在大型公共建筑中，为了满足使用延续性的需要，建筑需在使用的同时进行空间的灵活改动，结合建筑现有条件，通过利用本身或周边空间实现功能临时置换，体现空间的适应性。对于交通类建筑具有更高要求，不停航、不停运的空间改造是保证建筑功能正常运转的必要条件，在面临突发紧急情况需要功能调整时，也应能及时应对。广州白云机场 T1、T2 航站楼的不停航施工和平疫结合方面得到了体现。随着旅客流量的不断增长，T1 航站楼的安检服务满足不了使用要求，在对现状安检通道进行改造的过程中，为减少对现状航站楼布局的影响及保证旅客流线，改造利用安检场地后方资源空间，对功能空间进行临时置换，配合设施的调整，满足改造期间航站楼的总体安检使用需求；以疫情防控为代表的紧急功能需求下，打造灵活可变的隔离空间是主要防控措施之一，在 T2 航站楼的平疫结合改造设计中，探索了空间的适应性重组，充分利用现状楼内空间布局，对需增设功能的区域采用临时隔断进行分隔，方便施工与后期拆除，改造后的空间满足不同防疫检测需求，实现空间的可变与适应性。

广州亚运馆室外灰空间
Outdoor gray space of Guangzhou Asian Games Gymnasium

2.3 DYNAMIC AND RESILIENT SPACE

Space is the basic means for people to experience architecture and a major criteria to evaluate the quality of an architectural work. Through the organization of spaces, architecture provides a harmonious interpretation of the multi-dimensional principles of activity places. Modern architectural spaces have developed a freer way of organization, stronger public attributes and higher adaptability. Coherently organized, distinctive spaces offer users a rich, diverse space experience and manifest multi-dimensional essence.

2.3.1 Dynamic Spatial Organization

The exploration into space organization dates back long ago. At the beginning of the 19th century, French architect Jean-Nicolas-Louis Durand stripped off the details of architectural decoration and probed into the pattern of architectural space combination in his analysis of building typology. The space prototype diagram reviewed by him still works today. The organization method of architectural space differs as the function and type of buildings vary.

Today, with the continuous development of architectural connotation, the organization of architectural space has become more diverse. There are two evident types of space organization,

广州亚运馆历史展览馆可在赛后引入创意商业
Creative retails may be brought into the History Exhibition Hall of Guangzhou Asian Games Gymnasium after the games

i.e., one with logical process, and the other free and flexible. Large public buildings have complex functions. The combination of these two types of spatial organization can create building spaces with both logic and vitality.

1) Process-based Space Organization

Process spaces are constantly seen in building types marked by obvious process sequence, reflecting a certain degree of logic. In the architectural design of these types of buildings, attention should be paid to the creation of hierarchical spaces that show clear processes. For example, transportation buildings usually have an evident pattern of space organization. Airports have clear processes for check-in, security check, waiting, departure and arrival. The spatial design should, together with the supporting facilities, guide the continuous passenger traffic while considering their needs for short stays, so as to create diverse process space nodes. In the architectural space design of T2, Guangzhou Baiyun International Airport, dynamic and directional spatial organization is provided in response to the characteristics of passenger flow. Meanwhile, the project is planned with a generous, transparent and directional departure hall, a relatively less mobile security check, and a passenger waiting area with linear characteristics, so as to adapt to the characteristics of each process node and realize progressive organization of spaces.

2) Free space organization

Free spaces often appear in building types featured with abundant activities, reflecting a certain degree of freedom that is conducive to diversified process organization. The architectural design of these types of buildings should highlight the creation of free flowing multi-node spaces. This space typology can be seen in gymnasiums and performing arts centers. Guangzhou Asian Games Gymnasium, an early project of GDAD, creatively employs the design strategy of three-dimensional flowing space. The four venues of the Gymnasium are organized in parallel around a central space to create a continuous flow of heterogeneous spaces that may be used in several different sequences, and to develop a free-flowing space composition featuring an outer ring and partially accommodating processes. The organization of free flowing spaces around a core is also seen in Huadu Dongfeng Stadium and Zhaoqing New Area Sports Center. Such an approach adds the fun of exploration to the architectural spaces and fully reflects the freedom and vitality of free-flowing spaces.

2.3.2 Public Space Implantation

The importance of public space design for social vitality has long been self-evident. Jan Gehl once pointed out that human social activities especially depend on the setting of outdoor public spaces. The further opening of contemporary architectural space has given rise to a variety of public spaces inside buildings accompanied by new quality characteristics.

Public spaces inside buildings are conducive to communication similar to those outside. Both of them have the significance of stimulating vitality in public buildings. While outdoor public spaces are open in more aspects, indoor public spaces are open to a limited extent. In the design practice of some public buildings, the space interface may be moderately opened based on the use of basic functional spaces to foster favorable interior and outdoor public space environment and comprehensively enhance building vitality.

1) Outdoor public spaces

Outdoor public spaces provide places for people to enjoy daily life and public activities outside buildings, hence a high degree of publicity. A variety of outdoor pubic spaces can be provided leveraging the environment around buildings and being divided into public and private zones for the free selection of people. They can promote the crowd gathering and develop a vitality name card. For example, the Desheng Sports Center is composed of "three halls and one stadium", supporting commerce, parking and surrounding landscape space. Despite the limited land use conditions, we manage to design a number of outdoor public spaces such as outdoor activity platform, ground square, F2 platform, open-up floor and landscape roof. These public spaces marked by diverse forms, appropriate density, rich levels and great vitality are perfect choices for gathering/distribution, sports, stroll, leisure, display, publicity, commerce and other activities.

2) Indoor Public Spaces

Indoor public spaces, to some extent, have the characteristics of serving the public. In the meantime, their interaction with the main functions inside a building makes them just moderately public. As per the location relation with the building, they take the form of relatively independent ones and fully integrated ones. As people become increasingly aware of humanistic needs, the provision of indoor public spaces has become a significant way to enhance building vitality. In our bid proposal for Guangzhou West Tower, we create a highly public top floor observation hall that is not only connected with the interior of the building but also directly connected with the urban ground. With relative independence, the three-dimensional space also serves as a city parlor. In our design for T2, Guangzhou Baiyun International Airport, we create a partial public space that takes the form of a cultural square in the hall, differentiating it from the hall through the restriction of components and activating its diverse functions such as display, publicity, communication, etc.

2.3.3 Flexible Space adaption

The use of space should not be fully defined. Ludwig Mies van der Rohe pointed out that "the purpose of building services constantly changes, but we can't simply remove buildings". The Nakagin Capsule Tower presented by Kisho Kurokawa in 1972 brought about an ideological wave of organic building renewal. The adaptability of architectural space reveals both development potential and vitality.

The ability of architectural space to change with demand has become an important aspect of architectural space design. The diversity of space utilization demonstrates on one hand the plurality of functional uses and on the other hand the changeability and adaptability to the development of the times of architectural space. With constant socio- economic development, architectural space has developed a feature that can meet different use demands, which is conducive to building renewal.

1) Multi-purpose Space Utilization

Providing multi-functional spaces is an effective means to improve the service efficiency and adaptability of buildings and optimize the utilization of building resources. Multi-functional spaces can ensure a certain degree of flexibility in building use in case of any unexpected demands. In addition, we need to convert from the traditional concept of room classification to the concept of functional classification, allowing one same space to house different functions based on use characteristics and frequency differences. In sports buildings, sports venues should be multifunctional to accommodate both professional needs of the games and flexible use after games. For example, the spatial design of Guangzhou Asian Games Gymnasium allows all the sports venues to be converted to another type of venue as needed. The gymnastics hall can be flexibly changed into a basketball hall, and a base for mass sports that can operate independently. The Asian Games History Exhibition Hall can house creative retail after the Games except some retained exhibition functions. The exterior space for pedestrian distribution during the Games can be used as a public activity space after the Games. In real operation, the exterior gray space can support mixed uses and facilitate the organization of diversified events.

2) Space Resilience

In large public buildings, architectural space needs to be flexibly changed to ensure the continuity of use. Temporary replacement of functions can be achieved inside a large public building leveraging its architectural space or surrounding spaces in combination with the existing conditions of the building, which reflects the adaptability of architectural space. For transportation buildings, the requirements are higher. The renovation of space without suspending aviation services is essential to the normal operation of building functions and can guarantee timely response in the face of sudden emergencies that require functional adjustment. During the transformation of the security check channel in T1, Guangzhou Baiyun International Airport, we used the space behind the security check area as a temporary substitute to reduce the current layout and ensure passenger flow. During the renovation of T2 for uses both at ordinary times and during epidemic prevention, we explored the adaptive recombination of space and realized space changeability and adaptability by employing temporary partitions to separate the areas that need additional functions on top of the existing space layout in the building.

2.4 多元建筑形态与表皮建构

建筑形态与表皮是建筑的外在表现，创造了人们对建筑的初始印象。当代建筑设计中，线性与非线性作为两种建筑形态，各自具有鲜明的特征。当今建筑形态设计应基于功能要求和环境关系等，因地制宜选择合适的表达方式，并通过表皮系统丰富建筑表情，从而形成各具特色、与环境相互和谐的建筑作品。

2.4.1 多元共存的建筑形态

在 21 世纪之前，线性建筑形态作为传统类型被广泛应用，20 世纪 90 年代，美国建筑评论家查尔斯·詹克斯在《跳跃宇宙的建筑学》中将系统理论与建筑设计相结合，提出非线性建筑的观点[10]。随着数字化时代到来、科学技术开拓了新的设计与建造方法，推动了非线性建筑的蓬勃发展，如今非线性形态已与经典线性形态一起，成为建筑设计中的两种主要类型。

线性建筑与非线性建筑各有所长，在众多建筑实践中，均探索了这两种建筑形态类型。线性建筑以较低的技术水平要求、较强的操作性、秩序化的形态组织、方便建造施工等特征，在我国城市化建设中发挥了重要作用，在当今仍是主要的建筑形态类型之一，并随着时代的发展有了更丰富的内涵；非线性建筑以开放流动的形态、动态的时空特性、强大的形式生成潜力等优势，在日渐成熟的设计建造技术条件下，越来越多地出现在城市建设中，各种适度非线性、复杂非线性建筑陆续出现，呈现愈加多元的趋势。

1）线性建筑形态的当代呈现

线性建筑形态是从古至今使用时间最久、使用频率最高的经典建筑形态类型。通过理念与技术的融合，实现线性建筑形态语汇的打造，赋予建筑立面丰富的线性肌理，反映当代岭南建筑的形态表现形式。潭洲国际会展中心建筑造型设计从折纸艺术及广东剪纸取得灵感，通过翻折变化形成独特的线性屋面形态，并将呈“L”形布局的 5 个展馆、3 个连接厅、1 个登陆厅包络一体，形成连续起伏变化的造型；珠海横琴保利中心建筑外表皮百叶与内表皮玻璃幕墙共同构成适应地方气候的双层表皮系统，满足采光、遮阳和自然通风需求，长短离散渐变百叶与错位露台共同形成丰富多变的线性立面。主体建筑体形方正，100m × 100m 的建筑体量中心设有 40m × 40m 的天井，构筑“风之通道”，底层架空层与风的通道相结合，结合带巨型转换钢桁架的“框架—剪力墙”结构，实现建筑的“悬浮”效果，结构成就建筑，体现线性建筑之美。

2）非线性建筑形态的多元发展

非线性建筑形态具有有限度的非线性形态到复杂非线性形态等不同

潭洲国际会展中心线性造型与立面
Linear form and facade of Guangdong (Tanzhou) International Convention and Exhibition Center

珠海横琴保利中心线性造型与立面
Linear form and facade of The Poly Center in Hengqin, Zhuhai

广州白云机场 T1 航站楼简单非线性曲面造型
Concise, non-linear curved surface form of T1, Guangzhou Baiyun International Airport

广州白云机场 T2 航站楼有限度的非线性曲面造型
Limited non-linear curved surface form of T2, Guangzhou Baiyun International Airport

广州白云机场 T3 航站楼标准化建构逻辑的非线性曲面造型
Non-linear curved surface form with standardized construction logic of T3, Guangzhou Baiyun International Airport

广州亚运馆的复杂非线性曲面造型
Complex non-linear curved surface form of Guangzhou Asian Games Gymnasium

类型，程度不断发展。其流动形态与交通、体育类型紧密相关，通过标准化的建筑逻辑，构成看似复杂的建筑造型，使建筑形态延绵起伏，体现一种运动感。广州白云机场航站楼持续探索了从 T1 航站楼的简单非线性三维曲面造型、T2 航站楼的有限度非线性三维曲面造型，到 T3 航站楼标准化建构逻辑非线性曲面造型的变化，T1 航站楼的屋面采用三维曲面造型，将造型特征发挥得淋漓尽致；T2 航站楼的屋面具有适度非线性曲面造型特征，其连接楼和指廊以折面为主的局部弧面模拟非线性造型；T3 航站楼屋面在非线性曲面造型的基础上进一步做了逻辑化、标准化建构处理，组成基本曲面造型。在广州亚运馆中，以复杂的非线性形态、高标准的工程做法，开创了国内复杂非线性曲面造型建筑实践的先河，方案利用有机连续的金属屋面统领体操馆、综合馆和亚运博物馆等多个场馆，场馆犹如珍宝隐藏于屋面之下；场馆间、屋檐下的灰空间连贯舒展，大面积金属屋面轻盈飘逸，通过非线性设计，实现建筑从屋顶到平台的全方位控制，形成从表皮到流程的非线性效果。

汕头亚青会场馆丰富建筑形态下塑造的表皮效果
Building skin shaped by diversified architectural form of Shantou AYG Stadium

2.4.2 建筑形态的表皮建构

建筑表皮是建筑建构和艺术、文化的凝结，起着保护建筑本体和传递建筑情绪的作用。1851 年森佩尔在《建筑四要素》中暗示了表皮和结构的功能区别 [11]，1926 年柯布西耶提出现代建筑五大原则，将立面从结构体系中解放出来 [12]，此后，随着技术水平提升和“读图时代”的到来，建筑表皮逐渐成为聚焦公众视线的焦点，开始获得真正的独立。

如今表皮成为建筑造型的重要载体和建筑师表现个性的重要手段，在建筑师打造项目效果中起到了重要作用。表皮艺术的打造通过形态、肌理色彩和材料等方面的设计，共同呈现视觉、技术、文化、情感等多重内容，形态作为表皮基本关系，发挥主体效果的作用，肌理色彩和材料作为人性尺度的感知要素，使建筑表皮具有细腻的表现。三个要素共同促成表皮的整体表达。

1）形态塑造构成丰富界面

建筑表皮的形态塑造追求一种灵活的处理，结合地域的特征，形成独特的建筑形态，给建筑带来流动、畅快的视觉感受，不同的层叠进退，赋予建筑表皮灵活跳跃的丰富界面。在汕头亚青馆中，对建筑形态进行塑造，结合汕头的山海环境，将海浪、沙滩的形象转变成建筑形态，使建筑与场地呈现出协调丰富的效果，并结合亚青会的特性，展现年轻、活力、跳跃、灵动的特点，立面将海风与海浪的侵蚀感形成层层叠岩，刻画在建筑的立面、屋面及吊顶。建筑立面表皮与平面无严格的区分，通过不同层级的转换，平面的跌级可以转换到立面和吊顶，建筑以一种简洁的造型统领整体建筑，一气呵成、丰富流动。

珠海横琴保利中心“横琴风格”的非均质表皮
Non-homogeneous skin of The Poly Center in Hengqin, Zhuhai

2）肌理色彩形成整体效果

建筑表皮的肌理色彩形成了体验的重要元素，在公共建筑中，可通过平滑简约的肌理色彩，打造由内到外、从立面到屋顶的整体表皮造型，以朴素的色调打造统一的表皮效果，通过材料质感映衬环境，创造出与环境浑然一体的建筑，表现出建筑的简约大气，与环境相融合。汕头亚青馆基于山海的宏大环境，建筑立面选用金属幕墙、玻璃幕墙与石材三类材质，考虑基材尺寸及加工生产便利性与效果的综合因素，基本分格尺寸为统一的长方形，局部区域根据具体部位需求进行变化，通过横明竖隐玻璃幕墙系统，根据不同区域分级优化曲面分格，简化深化、生产、施工难度，室内外及屋面石材采用分格尺寸工字错缝铺砌；珠海横琴保利中心通过洁白光亮的色彩基调、光感素材及非均质表皮塑造，建构具有识别性的“横琴风格”表皮系统，映衬出阳光的灿烂、海空的湛蓝、绿意的盎然，在白色的衬托下，玻璃的青辉、金属的质感被真实还原，通过非均质肌理表皮——横向百叶形成的双层遮阳表皮系统，控制阳光的照射，百叶的穿插变化，形成丰富的立面形态。

3）材料选用丰富表现层次

建筑表皮的材料选择应考虑对建筑概念的回应，协同体现建筑的造型和肌理特点，实现表皮的特质。通过材料之间相互协调、相互映衬，适应于地域，呈现出独特的艺术感受，打造出浑然一体的建筑效果。汕头亚青馆选择简洁纯净的现代材料，考虑到地域气候的防风、防腐、节能等特点，亚青会场馆建筑立面主要采用不锈钢、玻璃幕墙及清水混凝土三种材料的组合形成建筑整体。不锈钢材料运用在体育场馆及会议中心的屋面、立面部分；玻璃幕墙运用在体育馆及会议中心的立面，采用隐索玻璃幕墙，干净大气；清水混凝土主要运用在

珠海横琴保利中心玻璃、金属的立面质感
Glass and metal facade of The Poly Center in Hengqin, Zhuhai

汕头亚青会场馆不锈钢、玻璃幕墙与清水混凝土的材料组合
Material combination of stainless steel, glass curtain wall and fair-faced concrete for Shantou AYG Stadium

室外可见结构柱、梁、板，观众看台，室外首层立面、室内主要空间、通道的外墙面。

2.4.3 建筑形态的意向表达

建筑形态的意向表达是体现建筑造型的重要方面。建筑参与到城市与自然环境的营造中，形成与环境相适应的形态，有利于体现整体感受。这种适应并非一种复原和模仿， 而是基于环境，做出抽象化、简洁化的形态处理。建筑形态的环境意向体现出了一定的艺术性、概念性特征，实现建筑与环境的协调。

在设计实践中，通过地景处理、自然空间设置等策略，实现了建筑形态与环境意向的和谐。通过建筑模拟自然的形象，打造与外界协调的室外空间，或模拟文化印象，形成一种从人到建筑的文化流动，通过形态的体现，来营造独特的地域环境效果，可分别从地景的拟态和环境景观渗透的方式，实现建筑环境的意向和谐。

1）地景意向回应周边环境

地景意向是建筑与环境之间的直接联系。建筑以一种开放的姿态面向环境，通过自由的建筑造型，将不同的建筑体量融为一体，环境的形象被抽象为有节奏的、连续跃动的建筑形态组合，最终凝固成具有旋律感的建筑曲线。亚青会场馆紧邻海岸，其形态灵感来源于浩瀚无垠的大海、奔腾不息的海浪和飞舞的浪花。自由流动的建筑与空间形态犹如大浪淘沙，表现出主题的意向。为使建筑更具海浪意向，更好地融入环境，方案采用了多种策略。首先对建筑体量进行了控制，最终将主体育场降低到 40m 以下，减小对周边环境的不利影响；其次打破了建筑体量，设置多种开放的“洞口”，体现出沙滩和海浪自由交织的通透感受；再通过多重室外平台将室内逐渐过渡到环境中，并用开放而又亲和的姿态来面对运动员和市民，人的参与及活动将成为建筑的有机组成，形成一种海边漫步的意向，带给使用者沉浸式的体验。亚青会场馆通过主题选取与形态控制，形成大浪淘沙、勇立潮头的精神气质。展现出潮汕人民敢为人先、

肇庆新区体育中心的“砚生水墨”意向
The imagery of "an ink painting of LOHAS Zhaoqing" for Zhaoqing New Area Sports Center

傲立潮头的拼搏精神，也寓意潮汕地区独具特色的海洋文化所兼具的开放性和包容性，实现了建筑和谐的环境意向。

2）衍射意向体现地域文化

衍射意向是建筑与地域文化之间的联系，体现出对文化的独特见解。为体现地域人文特征，建筑造型风格需要延续当地的建筑文化特质，形成一种灵动的建筑造型。肇庆新区体育中心便采用了“砚生水墨”这一主题概念，以回应典型的场地特质。方案总体布局呼应了河流等场地元素，力求使建筑设计、空间营造达到“水墨”之灵动境界，追求建设一个赛时集约高效，同时体现水墨人文气质的专业体育场馆。场馆间半室外空间向城市开放，构建一个聚集人气的城市客厅，为商业开发营造良好氛围。建筑与足球公园连成一体，通过南面的缺口，从体育场内可以观看到足球公园的景色。在极其有限的用地中，沿城市道路一侧布置场馆建筑，靠河岸一侧用地则作为足球公园使用。端砚形态丰富，其所生成的水墨造就了中国传统书法与绘画。建筑采用玻璃幕墙与金属铝板相结合，“V”形柱与曲面屋面相结合，形成场馆合一、契合山水地貌的流畅优雅体形。以“水墨星空”为概念的建筑泛光设计，使自由形态的建筑与大自然的星空、蜿蜒流淌的水岸相融合，如端砚般质刚，似水墨般柔美，营造出水墨星河之灵动的视觉意向。

2.4 DIVERSIFIED BUILDING FORMS AND SKIN CONSTRUCTION

Architectural forms and skins are the external appearance of buildings, determining people's first impression of them. In contemporary architectural design, linearity and non-linearity are two architectural forms with distinct characteristics. Nowadays, the architectural form design should be based on functional requirements and relationships with the environment. Also appropriate expression approaches and skin systems should be used to create distinctive buildings that are in harmony with the environment.

2.4.1 Diverse Architectural Forms

Before 2000, linear architectural form was widely used as a traditional style. In the 1990s, American architectural critic Charles Jencks combined system theory with architectural design in *The Architecture of the Jumping Universe* and proposed nonlinear architecture. With the advent of the digital age, new design and construction approaches powered by science and technology have promoted the vigorous development of nonlinear buildings, and now nonlinear form has become a fashionable style in architectural design together with the classic linear one.

Linear and nonlinear buildings have their own strengths. Linear buildings have played a critical role in China's urbanization due to strong operability, and remain today one of the main architectural forms. Non-linear buildings boasting open and flowing forms, dynamic space-time characteristics and great potential to generate forms, have been more and more commonly seen in urban construction as the design and construction technologies mature.

1) Presentation of Linear Architectural Form in Modern Days

Linear architectural form is the classic form that has been most frequently used since ancient times. Through integration of concepts and technologies, the formal vocabulary of linear architectural is created to show the expressional form of modern Lingnan architecture. The architectural form of Tanzhou International Convention and Exhibition Center is inspired by the art of paper folding and Guangdong paper cutting, featuring a continuous undulating appearance through folding and changes. The building of The Poly Center in Hengqin, Zhuhai boasts a double-skin system with louvers as the external skin and glass curtain wall as the internal skin to adapt to the local climate. The long and short gradient louvers and staggered terraces add diversity to the linear facade.

2) Diversified Development of Nonlinear Architectural Form

Nonlinear architectural forms range from those with limited nonlinearity to complex ones. The flowing forms are closely related to transportation and sports, as the undulating building forms remind people of motion. The seemingly complex architectural shapes are formed based on standardized architectural logic. The terminals of Guangzhou Baiyun Airport continuously explore the changes in nonlinearity from simple nonlinear three-dimensional surface (T1), to limited nonlinear three-dimensional surface (T2), and then to standardized logical nonlinear surface (T3). The design of Guangzhou Asian Games Gymnasium has set up a precedent for complex nonlinear curved surface in China, which enables all-around control from the roof to the platform and nonlinearity from the skin to the process.

2.4.2 Skin Construction of Architectural Forms

The architectural skin is the result of the integration of architectural construction with art and culture, and plays the roles of protecting the architectural body and conveying architectural emotions. In 1851, Semper hinted at the functional difference between skin and structure in *Die vier Elemente der Baukunst* (Four Elements of Architecture). In 1926, Le Corbusier proposed five points of modern architecture, one of which separated the facade from the structural system and gradually freed the design of building skin.

Now the skin has become an important carrier of architectural forms and an important means for architects to express their individuality. The design of skin involves form, texture, color and material, which jointly present vision, technology, culture and emotion. In addition, texture color and material as the perceptual elements of scale of humanity, refine the expression of building skin.

1) Dynamic Interfaces Shaped by Architectural Forms

The forms of building skins value flexibility. Unique architectural forms incorporating local characteristics present a smooth and lively visual experience. Abundant skin forms characterized by youth and vitality are used in the design of Shantou Asian Youth Games Stadium. There is no strict distinction between the facade skin and floor planes of the building. Concise design approach and lively lines portray a coherent and spectacular building.

2) Overall Effect Delivered via Texture and Color

Texture and colors of the building skins are important elements of experience. For public buildings, smooth and simple texture and colors may work with clean tones to define consistent skin, and create buildings fully blended into the environment. Shantou Asian Youth Games Stadium uses three facade materials, namely metal curtain walls, glass curtain walls and stones. The facade is basically divided into rectangles of the same size, except for some curved surfaces optimized as needed in local areas. Poly Center in Hengqin, Zhuhai adopts white and bright color tone, sensitive materials and non-homogeneous skin to constitute the iconic skin system of Hengqin style and diverse facade forms.

3) Architectural Expression Enriched by Materials

The material selection of the building skins should respond to the architectural concept, reflect the characteristics of building form and texture, and express the properties of the skin. The materials should be coordinated and contrast with each other, adapt to the local conditions, and present unique artistic quality to create a seamless and consistent architectural effect. Shantou Asian Youth Games Stadium uses simple and clean modern materials. Specifically, the building facade adopts a combination of three materials: stainless steel curtain wall, glass curtain wall and fair-faced concrete. Stainless steel and glass curtain walls are used on the roof and facade, which appear clean and magnificent; fair-faced concrete is mainly used for the exterior visible structures and the main interior space, which is concise and clear.

2.4.3 Intent Expression of Architectural Forms

The relationship between architectural form and environmental intent is an important aspect reflecting the temperament of a building. Buildings are engaged in creating urban and natural environments in forms that adapt to the environments. Such adaption is not about restoration or imitation; instead, it is an abstract and concise form developed from environments.

In design practice, the harmony between architectural form and environmental intent is achieved through strategies such as landscape treatment and natural space creation. Buildings simulate the image of nature to create an exterior space that is in harmony with the outside world, or simulate cultural impressions to enable a cultural flow from people to buildings. Then architectural forms are utilized to reflect the unique regional environment.

1) Landscape Intent in Response to Surrounding Environment

Landscape intent is the direct link between buildings and the environment. The building opens to the environment, and integrates with it through free architectural forms. By abstracting the image of the environment, a combination of rhythmic, continuous and dynamic architectural forms is created and turned into melody-like building curves. The form of the Asian Youth Games Stadium is inspired by the boundless sea. In order to better integrate the building into the environment, various strategies are adopted in the design. Firstly, the building volume is properly controlled to reduce the adverse impact on the surrounding environment; secondly, the building volume is divided into smaller ones by a variety of open and transparent "holes", bringing a transparent experience coupled with the freely interwoven beach and waves; last but not least, several levels of exterior platforms are planned to gradually lead the interior to the exterior, implying a walk by the sea. The stadium achieves the harmony between architectural form and environmental intent through theme selection and form control.

2) Symbolic Intent Echoing Regional Culture

Symbolic intent is the connection between buildings and regional culture. It reflects a unique insight into culture. To showcase the regional humanistic characteristics, the architectural styles should carry on the local architectural cultural features. For example, Zhaoqing New Area Sports Center adopts the concept of inkstone as a response to the cultural features typical of Zhaoqing. The overall layout of the Sports Center responds to elements such as rivers, while architecture and space try to convey the same grace and elegance as depicted in Chinese calligraphy and painting. The floodlighting design blends the free-flowing building form with the starry sky and the meandering river banks, vividly portraying a sports facility rising beside the gently flowing starry river.

2.5 生态融合与有机平衡

生态环境是与人类相关的各种自然元素的总和。建筑作为联系人类与自然之间的物质载体，其衍生与发展有其固有的生态属性。岭南建筑在湿热多雨的气候条件下形成了对自然开放、与生态交融的独特风格，追求建筑与生态的和谐共生。随着社会生态意识的提升，当代岭南建筑更加注重绿色环保，特别是对于大型公共建筑，往往需要大量的能耗才能维持其正常的运行，因此，基于大型公建的绿色技术创新，对推动建筑可持续发展，维护生态平衡具有重大而长远的意义。

2.5.1 “自由呼吸”建筑

呼吸是生命体的重要特征之一，会呼吸的建筑注重利用自然通风等技术改善热环境，创造舒适微气候，实现节能减排。梁思成建筑奖首位外籍设计师获得者——马来西亚建筑师杨经文将创造自然通风条件融入其建筑实践，他提出“在生态结构上包括了生命体和无生命体，在设计中要尽量把生命体和无生命体进行有机的统一”，自然风作为生态结构中流动的有机体，能为建筑带来活力与生命力。随着社会经济水平与科学技术的发展，当代公共建筑体量不断增大，采暖制冷能耗成为建筑总体能耗的主要部分。岭南地区的建筑在常年湿热的气候背景下，需要重点关注减少太阳热辐射、降低空调制冷能耗，营造鲜活清爽的空间气息，自然通风的利用将提供一种有效的解决思路。

1）双层呼吸表皮

建筑表皮作为分隔建筑室内外的界面，对建筑能耗管理起到关键作

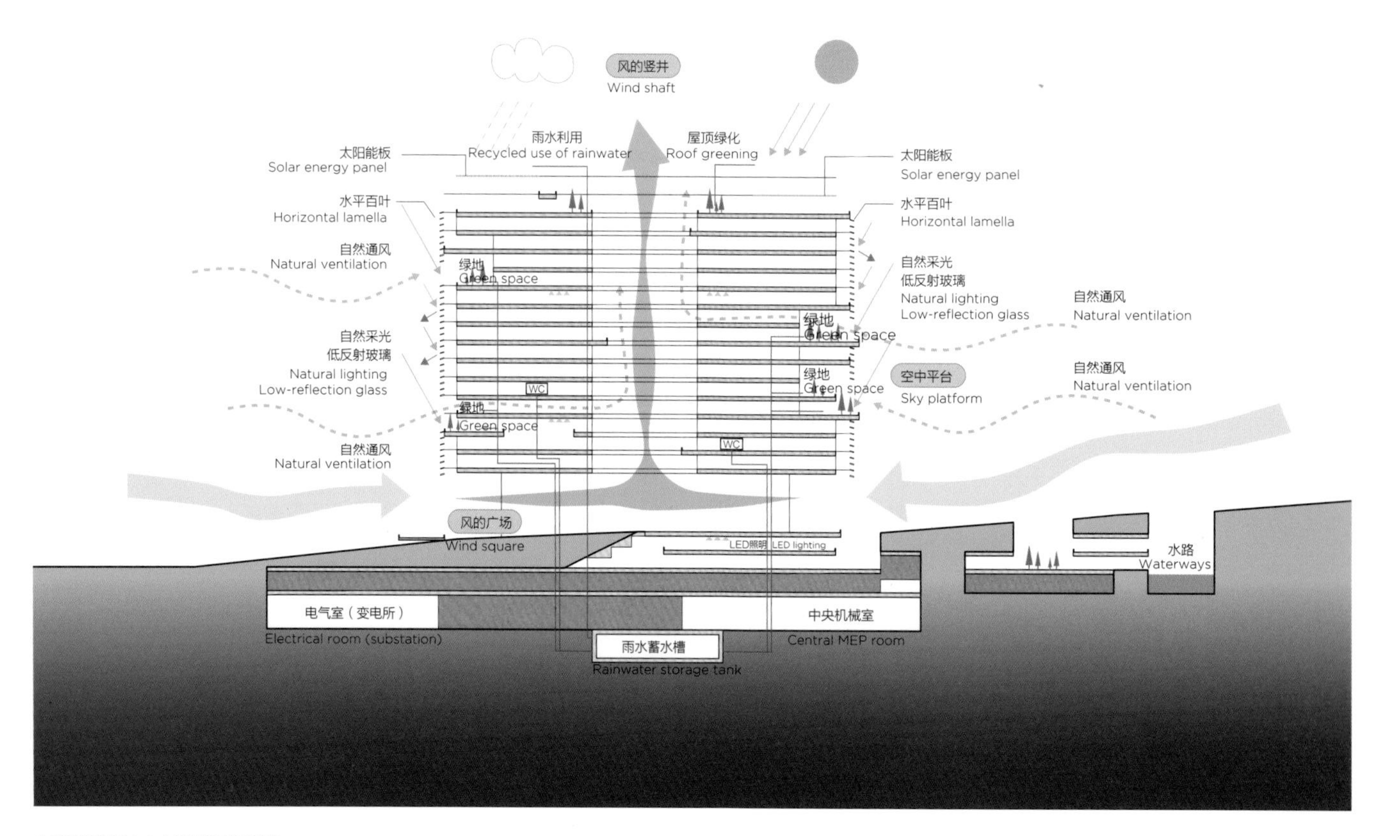

珠海横琴保利中心立体呼吸系统建构
3D breathing system of The Poly Center in Hengqin, Zhuhai

珠海横琴保利中心双层呼吸表皮
Breathable double skins of The Poly Center in Hengqin, Zhuhai

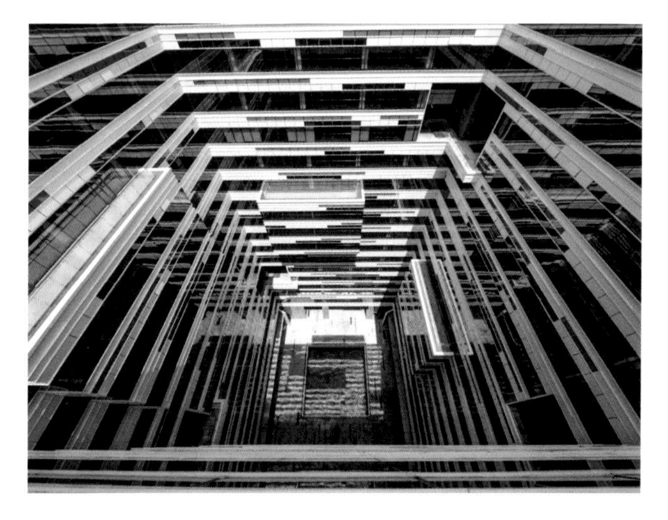

珠海横琴保利中心立体风之通道
Vertical air channel of The Poly Center in Hengqin, Zhuhai

用。传统岭南建筑常通过青砖等具有良好热阻性能的材料作为外围护结构，有效隔绝热传导，保证室内热舒适度。当代岭南建筑将玻璃幕墙作为常用的表皮材料，能实现开阔通透的视觉效果，保证室内采光。由于玻璃幕墙在保温隔热性能方面具有一定的局限性，其外围可结合其他外材料形成多重表皮，在外表皮与玻璃幕墙表皮之间形成缓冲层，减少太阳辐射，增加空气流通。在珠海横琴保利中心项目中，通过构建内玻璃幕墙表皮与外遮阳百叶表皮，实现了采光、遮阳和自然通风三种功能结合，共同组成会呼吸的立体生态建筑表皮，降低建筑能耗。

2）生态呼吸系统

自然通风能为建筑室内提供新鲜的空气，平衡室内温度，为建筑本体“过滤”杂质，呈现出绿色低碳、清新可近的特质，营造高品质室内环境。生物内部与外界环境进行气体交换需要依靠呼吸系统来完成，对于建筑而言，同样需要通过相关的“呼吸系统”进行实现。珠海横琴保利中心主体建筑体形方正，正南北朝向，造型与功能相适应，项目总高度约 100 米，建筑塔楼标准层平面尺寸约 100m × 100 米，建筑底部通敞，主体结构从三层开始向外悬挑约 12.5~15.5 米，架空层与露天天井共同形成“风之通道”，清新自然的风沿室外绿丘徐徐而上，贯穿塔楼中央，进入各主要功能房间内部，营造可呼吸的办公空间。

2.5.2 立体绿化嵌入

绿化植物对人工环境能起到良好的改善作用，提升建筑环境的生态效果，是实现人工环境自然化、优化城市生态的有效方法。随着技术的进步及绿色理念的推广，绿化的设置已成为建筑环境中不可或缺的内容，当代建筑与绿化的结合形式不断发展，内涵不断增加，产生了从平面绿化到立体绿化的丰富变化。

在岭南地区湿热的气候条件下，将绿化立体引进建筑当中不但能提升空间品质，同时能营造舒适的微气候。大型绿化庭院和立体绿化是两个显著的方式，可以起到营造景观和改善环境的双重作用，通过绿化庭院的嵌入，能形成体验自然的场所；通过地面、墙面、屋顶等立体绿化的点缀，为人们提供更多与生态环境交流的契机，提升自然体验。

1）大型绿化庭院

绿化庭院是提升建筑自然环境的显著方式。传统岭南建筑常通过房子的围合形成一个个“院”，在院中营造园林景观，形成“自然的容器”，人们可以在其中感受自然的气息。在岭南地区大型机场航站楼中，通过绿化庭院的嵌入，能有效提升室内景观效果及空气品质。广州白云机场 T2 航站楼率先在枢纽航站楼内营造充满岭南地域特色的大型垂直绿化庭院——岭南花园，成为航站楼内的中央绿肺，也让旅客在现代化的航站楼内可以感受到传统岭南园林的魅力，留下美好的出行体验。

2）多维绿化叠加

多维的绿化可以形成较好的环境生态点缀。通过充分利用场地斜坡、台阶，以及建筑的墙面、屋面等立体维度，引入立体绿化，创造良好的建筑空间品质，同时利用绿化的蒸腾与反射等作用，营造舒适的微气候。在珠海横琴保利中心项目中，露台绿化、屋面绿化、绿化广场等交相呼应，组成多维立体绿化系统。带状与点状立体绿化露台镶嵌于建筑表皮之上，与室内办公空间形成不同的组合，提供了绿色生态、舒适趣味的交流空间；屋面绿化营造了亲近自然、开阔共享的立体交流平台；与地下室功能有机融合的室外绿化广场，生态空间交错叠合，立体绿化意趣盎然。

2.5.3 生态有机适应

有机适应生态，不仅体现为建筑形态与环境的融合，还体现为建筑对环境的主动适应与自我调节能力。“有机”的概念最早由美国建筑师路易斯·沙利文从生物学领域借用、移植而来。“有机建筑”则是由美国建筑师弗兰克·劳埃德·莱特提出，莱特认为：“有机建筑表现出对自然环境的亲和，如同一个有生命的有机体”。早期对有机建筑的探索更多倾向于建筑形态的呈现，随着时代的发展，建筑的有机层面有更广泛的延伸。1963 年，奥戈雅出版专著《设计结合气候：建筑地方主义的生物气候研究》，确立生物气候地方主义理论，将建筑的有机与生物的自我调节特性进行关联[13]。

当代“双碳”理念的提倡对岭南建筑的生态适应性提出了更高的要求，通过创新绿色能源利用及气候适应性技术手段，建立建筑自身的生

广州白云机场 T2 航站楼岭南花园大型绿化庭院
Green courtyard of Lingnan Garden in T2, Guangzhou Baiyun International Airport

物气候调节机制，主动适应岭南气候特征，打造舒适怡人的建筑环境，力求创造节能、健康、舒适的绿色建筑。

1）可再生能源有机利用

可再生能源包含太阳能、风能、生物质能等非化石能源，可再生能源的利用伴随着人类长期的历史进程，其具有储藏量大、清洁环保等优点，在当代提倡可持续发展理念的背景下，对于能源结构改善、生态环境保护具有重要意义，机场航站楼建筑体量大，且往往存在大进深的室内空间，自然采光受限。在广州白云机场 T2 航站楼中，通过“长大带形天窗 + 渐变旋转式吊顶”采光设计，直接利用太阳光补充室内照度，解决办票大厅与值机大厅等大进深空间的采光和防眩光问题，减少人工照明能耗。结合航站楼庞大的平面布局，率先在航站楼屋面上设置一定规模的分布式光伏发电系统，将光能转化为电能，项目自2018 年运行以来，平均年度系统实测发电量超过 200 万度。通过直接或间接利用可再生能源，为航站楼高效环保供能。

2）被动节能技术有机整合

被动节能技术是指通过非机械电气设备干预而实现降低建筑能耗的技术手段，如建筑遮阳设置、外围护保温隔热等技术。机场航站楼因外围护结构面积巨大，其热工性能的提升对于节能减排具有显著的效果。在广州白云机场 T2 航站楼中，首次在航站楼上采用被动防热节能围护构造技术，幕墙门窗采用绝热型材配置高透绝热玻璃，非透明金属屋面采用高反射率的反射隔热涂料，内嵌保温材料，实现夏季隔热，冬季保温的动态调节。西向区域采用与玻璃幕墙结合的可调节电动遮阳百叶系统，百叶可根据不同日照高度进行角调节，保证西向室内候机空间的热舒适度，大大降低空调能耗。

珠海横琴保利中心多维立体绿化
Multi-level greening of The Poly Center in Hengqin, Zhuhai

3）设备改造促使能效有机提升

设备系统是建筑体系的重要组成部分，也是建筑中能源消耗的主体，其能效的提升对于建筑节能具有重要的意义。相对于建筑本体，设备系统的生命周期更为短暂，需要更频繁的改造更新，因此可充分利用其改造更新的契机，引入更高能效的系统。在广州白云机场 T1 航站楼中，基于设备系统的后评估，对电气、暖通空调、给水排水进行有针对性的改造，将能源利用率不高以及老化的设备进行更新，提升设备系统整体能效，助力白云机场打造绿色低碳航站楼。

广州白云机场 T2 航站楼“大屋檐 + 可调电动百叶综合外遮阳”系统
Crown cornice + adjustable electric louver external sunshade system of T2, Guangzhou Baiyun International Airport

广州白云机场 T2 航站楼旋转渐变天花引入均匀自然光线
Evenly-distributed daylight introduced by the rotating ceiling of T2, Guangzhou Baiyun International Airport

2.5 ECOLOGICAL INTEGRATION AND ORGANIC BALANCE

The ecological environment is the aggregation of all natural elements associated with human beings. The extraction of natural elements and the simulation of natural forms, textures and colors in terms of building shape and components or materials using new technologies can inspire people's wonderful imagination about the ecological environment. Green technology innovation based on large public buildings is of great, long-term significance to sustainable development of buildings and ecological balance.

2.5.1 Breathable Architecture

Breathable architecture focuses on the improvement of the thermal environment using natural ventilation technology, which can foster a comfortable microclimate and reduce energy consumption and emission. Malaysian architect Kenneth King Mun YEANG integrates the creation of natural ventilation conditions into his architectural practice, advocating that life and inanimate entities should be organically unified as much as possible in design and that natural wind can bring vigor and vitality to buildings.

Due to an increasingly large volume of contemporary public buildings, heating and cooling have become the main energy consumers of buildings. Given the humid and hot climate of Lingnan, architects should focus on how to reduce solar heat radiation, cut cooling energy consumption, and foster a refreshing space atmosphere. The utilization of natural ventilation will provide a fruitful solution.

1) Breathable Double Skin

As the interface separating the interior and exterior of buildings, building skin plays a key role in the management of building energy consumption. Modern Lingnan architecture takes glass curtain wall as a common skin material, which can achieve an open and transparent visual effect and ensure interior daylighting. The periphery of the glass curtain wall may be combined with other external materials to form multiple skins thus a buffer layer in between to reduce solar radiation and increase air circulation. In The Poly Center in Hengqin, Zhuhai, an inner glass curtain wall skin and an outer sun shading shutter skin are built to integrate daylighting, sun shading and natural ventilation functions and form a three-dimensional respiratory ecological building skin that reduces building energy consumption.

2) Ecological Breathing System

Natural ventilation can supply fresh air, balance temperature and "filter" impurities inside a building, thus develop green, low-carbon, refreshing and approachable characteristics and foster a high-quality interior environment for the building. The main building of The Poly Center in Hengqin, Zhuhai Center has an open base, and its main structure overhangs about 12.5-15.5 m outward starting from F3. The open-up floor and open patio form a "wind passage" for refreshing natural wind to slowly rise along the exterior green mounds, pass through the center of the tower and enter main functional rooms to create breathable office spaces.

2.5.2 Vertical Greening Implantation

Greening plants can improve the artificial environment and elevate the ecological effect of the built environment. They can effectively naturalize the artificial environment and improve the urban ecosystem. With the progress of technology and the promotion of green concepts, greening has become an indispensable part of the built environment. The constant evolution of architecture-greening combination form and meaning has resulted in diverse changes from planar greening to vertical greening.

Given the humid and hot climate of Lingnan, introducing vertical greening into buildings can improve space quality and foster a cozy microclimate. Large green courtyard and vertical greening are two significant ways to create a landscape and improve the environment. The embedded green courtyards can offer places for people to embrace nature, while the vertical greening on the ground, wall and roof bring more opportunities for people to communicate with the ecological environment and enjoy nature.

1) Large Green Courtyards

Green courtyards can evidently improve the natural environment of buildings. Traditional Lingnan architecture usually features courtyards with garden landscapes enclosed by houses. These "natural containers" offer places where people can soak up the charm of nature. For large airport terminals in Lingnan region, the implantation of green courtyard can effectively upgrade interior landscape effect and air quality. Among them, T2, Guangzhou Baiyun International Airport is the first to create a large vertical green courtyard with Lingnan characteristics, the Lingnan Garden, inside a hub terminal. As the central green lung of T2, Lingnan Garden invites passengers to get a dose of attractive traditional Lingnan gardens in a modern terminal, adding fond memory to their travel experience.

2) Multi-level Greening Superposition

Multi-level greening can impressively embellish a built environment from the ecological perspective. It may be created in combination with site slopes, steps, building walls and roofs to develop quality architectural spaces. Meanwhile, the transpiration and reflection of greening can also foster a pleasant microclimate. Zhuhai Poly Center is an example of a multi-level greening system which integrates terrace, roof and square greening. Its strip- and dot-shaped green terraces on various levels are inlaid in the building skin, forming different combinations with interior office spaces and creating green, comfortable, interesting spaces for communication. Its roof greening offers an open, shared 3D communication platform that is close to nature, while its exterior green square organically integrated with basement functions gives rise to interesting interlaced and overlapped ecological spaces.

2.5.3 Organic Adaption to Ecology

Organic adaption to ecology reflects a building's capability of active adaption and self-adjustment in response to environment changes. The early exploration of organic architecture was more inclined to the representation of architectural forms. With the development of the times, the organic level of architecture has been extended more widely. In 1963, Victor Olgyay established the theory of bioclimatic

regionalism in his monograph *Design with Climate: Bioclimatic Approach to Architectural Regionalism*, linking the organic nature of architecture with the self-regulation characteristics of biology.

The contemporary advocate of "carbon peak and carbon neutrality" puts forward higher requirements for the ecological adaptability of Lingnan architecture. Innovative green energy utilization and climate adaption technologies can help establish a building's own biological climate adjustment mechanism and allow it to actively adapt to the climate characteristics of Lingnan. That is how energy-saving, healthy and comfortable green buildings are created.

1) Organic Utilization of Renewable Energy Sources

The use of renewable energy is accompanied by the long-term historical process of mankind. With such merits as large reserve, cleanness and environmental friendliness, it is of great significance to the improvement of the energy structure and the protection of the ecological environment. The design of T2, Guangzhou Baiyun International Airport adopts "long strip skylight + gradual rotation suspended ceiling" for daylighting, and directly use sunlight to supplement interior illumination, which greatly cuts the energy consumption of artificial lighting. The project is also the first to install a distributed photovoltaic power generation system on the roof of the terminal, which can produce over 2 million kWh electricity each year on average according to actual measurement. The terminal is supplied with efficient, environment-friendly energy through direct or indirect use of renewable energy sources.

2) Organic Integration of Passive Energy-saving Technologies

Passive energy-saving technologies refer to the technological means that reduce building energy consumption through the intervention of non-mechanical & electrical equipment, such as building sun shading, envelope insulation, etc. For buildings with an enormous envelope structure such as airport terminals, the improvement of thermal performance can significantly reduce energy consumption and emission. T2, Guangzhou Baiyun International Airport is the first terminal that employs passive thermal protection and energy-saving envelope construction technology, using ultra-clear thermal insulation glass and thermal insulation profiles for curtain wall doors and windows and high reflectivity thermal insulation paint and thermal insulation embedment materials for the nontransparent metal roof to achieve dynamic adjustment of thermal insulation in summer and winter.

3) Organic Improvement of Energy Efficiency through Building Service Renovation

The building service system is part and parcel of the building system and consumes the most energy inside buildings. Therefore, improving the energy efficiency of the building service system is of great significance to building energy conservation. Compared with the building itself, the building service system has a shorter life cycle and thus requires more frequent renovation and renewal, which could be a good opportunity for the building to have a more energy-efficient system. T1, Guangzhou Baiyun International Airport, based on the post-evaluation of the building service system, renovates its electrical, HVAC, and plumbing/drainage systems and replaces all equipment that is either aged or has low energy utilization, so as to improve the overall energy efficiency of the building service system and create a green, low-carbon terminal.

2.6 人文介入与体验感知

人文是地域性建筑的一个重要特征，其体现出建筑所在地的自然、社会发展状态，强调人性化的体验。介入和感知人文，即从当地环境分析入手、挖掘可作为设计启发的文脉，将其转化为具有文化品质的设计内容。以文脉转译结合人的活动，打造多重交往场所，以人文场景提升建筑品质，传承当代建筑的“人本”精神。

2.6.1 形象的文脉元素

文脉是文化元素的重要组成部分，其指出了文化要素之间以及文化与环境之间的约定关系。从罗伯特 · 文丘里首先将“文脉”的概念引入现代建筑领域至今 [14]，这个概念已成为建筑实践与研究的热门课题，其强调建筑是整体环境的一部分，讲究视觉和心理上的连续性。在建筑、城市、景观等领域的工程实践中，文脉的概念已经得到广泛应用，成为当前人居环境学科理论中的一个基本内容。

岭南地区独特的文化特质，孕育出了具有鲜明地域特色的建筑，其体现出历史文化价值及朴素而巧妙的营建智慧，为当代岭南地区建筑设计的文脉源泉。在设计实践中结合文脉特征，使现代建筑与地域传统相联系，促进传统文化的传承和发展。岭南地区包容并蓄的人文遗产为建筑创作提供了宝贵的人文积淀，也为探究文脉、发掘价值提供了驱动力。

1）文脉记忆探究

随着全球化的发展，不同国家和地区的建筑风格相互影响，趋于同质化，需要通过对传统文脉的传承，保持建筑文化的多样性。全球化进程中的当代岭南建筑应该兼具本土精神与时代特色，通过文脉记忆探究，保持地方性语言的表达。在广州白云机场 T3 航站楼前的 GTC 中，对城市历史文脉进行了深入考究，GTC 造型设计上通

广州白云机场 T3 航站楼前 GTC 体块设计体现对历史文脉的探究
The building block of GTC in front of T3, Guangzhou Baiyun International Airport reflecting a probe into historical and cultural context

过中间连接航站楼的体块与左右两侧六个停车楼体块，象征五羊传说中仙人赠予广州优良“稻穗”——“一茎六穗”的故事。在 GTC 的室内设计中，延续岭南故乡水的历史文脉，营造出历史与现代结合的“新故乡水”景观，层层绿化，跌水瀑布，打造具有岭南韵味的羊城新客厅。

2）文化符号转译

“转译”最先是语言学中的一种翻译行为，如今其概念已拓展到建筑的领域，文化符号的转译意味着对文化传统符号的表意系统转换到新的建筑系统之中，保持意义不变，这对于建筑地域文化的表达具有重要的指导意义。广州白云机场 T2 航站楼探索了传统岭南庭院与建筑的当代转译，空间上结合旅客流程，在航站楼内打造一处岭南花园，给旅客带来当代岭南庭院新体验；形态上抽象出传统岭南建筑鳞次栉比的屋顶特征，将花园中垂直绿化墙面顶部设计成自由起伏的折线，产生灵动轻盈的岭南建筑新形态；材质上通过模拟岭南建筑的特色肌理，形成清水文化砖、“U”形玻璃与彩釉玻璃三种表皮类型，穿插覆盖于绿化墙面，呈现出传统岭南建筑的现代表皮意向。南粤古驿道梅岭驿站项目对乡土建筑语境与传统建筑元素进行转译，首先通过解构当地民居，提取出坡屋顶式的三角体块、均衡体量、传统材料三个原型特征；其次对屋顶三角模块进行形态模拟，生成正方形单元，并以多样的组合方式构成大的复杂屋面体系，叠级上升顺应场地坡度；再而运用当地传统木材、青砖、灰瓦等材料，以现代建构技术进行搭建，创造出朴素自然的驿站形象，与当地传统建筑一脉相承；最后对当地居民灰空间公共活动的传统场景进行提取，打造檐下活动空间，形成对传统生活场景的再现，实现意境重构。

南粤古驿道梅岭驿站转译呈现的丰富细部
Rich details translated and represented by Meiling Station of South China Historical Trail

深圳机场新航站区 GTC 内部整合多元业态
Integration of diverse trades inside the GTC in the new terminal area of Shenzhen Bao'an International Airport

广州猎桥变电站设计了丰富的城市公共场所
Diverse urban public places in the 110kV Lieqiao Substation, Guangzhou

2.6.2 聚集人气的交往场所

交往场所广泛出现于各类公共建筑中。伊东丰雄认为公共建筑“聚集了很多人”比具体功能更有意义，扬·盖尔在《交往与空间》中也提出交往场所对公共空间的活力起着重要作用。交往场所不仅有利于建筑功能多元化发展，还对城市人文活力大有裨益[8]。

关注城市公共空间的聚集和交往属性，致力于提升建筑场所的“人情味”，在当前设计实践背景中具有积极意义。场所的出现伴随着身份认同、情感归属、心理安全等方面的考量，对交往场所进行打造，应以空间为基础，探索人们交往活动的方式，提升空间的历史和文化积淀。

1）活力交织

在交通建筑中，通过不断挖掘交通功能之外的附加服务，赋予场所更多的可能性，使其由通过地转变成目的地，吸引人流聚集共享，打造舒适活力的场所。在城市工业建筑中，除了满足建筑自身功能需求的同时，通过融入多元的城市环境之中，在公共性方面释放出多维的价值，成为充满期待的城市活力点。深圳机场新航站区 GTC 是一个连接 T3 航站楼与航空城、轨道交通的多元化交通核心，集地铁、大巴、出租车、社会车辆等多种交通方式为一体。GTC 突破传统交通中心功能，整合了丰富的商业与餐饮，提供了多样的附加服务，并通过平台与连桥与周边毗邻的商务酒店、公寓、办公等建筑相连通；作为“拉近市政设施与市民距离”的首个落成项目，广州猎桥变电站超越了传统变电站的概念，为了改善变电站空间消极的城市问题，方案置入科普馆、展厅、花园等多重内容，设有与江景、城市交融的特色景观，通过注入惬意自然的日常生活场景，让原本冰冷邻避的工业电站融入多元的城市空间之中。变电站不再仅是城市的供电装置，还是一座化解市民疑虑的体验式科普中心，更是一处被精心设计的小型城市公共

广州白云机场 T2 航站楼内人性化场所设计
People-oriented places inside T2, Guangzhou Baiyun International Airport

场所，使得原本封闭消极的变电站在城市公共性方面释放出多维的价值，成为充满期待的城市活力点。

2）人性化设计

结合人们的生理特征、心理因素、行为习惯等层面，通过精心设计，提升场所品质，促进人与人、人与建筑、人与城市之间更好的交往互动，让人们感受到自己是被尊重的个体，体现人文关怀。广州白云机场 T2 航站楼旅客候机区设置母婴室、儿童活动区、文化展示区，柜台及座椅结合人体工程学设计，并满足无障碍使用需求，座椅扶手设有充电装置。通过精细化设计、人体工程学设计和无障碍设计，给各类人群提供无微不至的关怀；广州猎桥变电站将人文关怀植入工业建筑，使其由“邻避设施”变成城市人文休闲节点。建筑采用谦逊圆润的形体，与周边城市地标取得相互协调的舒适效果，又以理性克制的模数化立面体现工业建筑的应有特色，穿孔铝板的外表皮赋予了建筑干净轻盈的视觉效果，配合灵动变化的泛光设计，呈现出“月光宝盒”般独特的艺术效果，为广州珠江夜景新添绝美的人文打卡圣地。

2.6.3 情景静态主题与动态交互

人们并非仅仅依赖视觉来感知空间和获得心理上的触动，将氛围情景设计引入空间构图，丰富着人们的场景感知。卒姆托曾提到：“事实上，任何一个成功的建筑寻求的都是对人的综合知觉的传达，这种知觉综合性恰恰是超越于单纯的视觉欣赏，而依靠于人在空间中的感受和体验。”在进行建筑设计时亦应关注人们的综合性情景体验。

情景主题是基于建筑空间建构后的环境氛围营造，其不仅是一种视觉特征，更能触动人们各种感觉——重量、温度、光泽、粗糙或细腻的表面、甚至是气味，形成一种综合感知。在公共建筑中，使用情景主题与人性化的设计，使建筑的情景承载丰富的体验，调动人体综合感

深圳宝安机场卫星厅丰富的空间主题
Diversified space theme inside Satellite Hall, Shenzhen Bao'an International Airport

官知觉，带来建筑情景感知的整体印象，提升建筑的人文韵味。

1）静态情景主题

在静态情景主题的表现中，通过对自然、人文等多主题打造，形成一种整体性的建筑情景氛围。深圳机场卫星厅运用多种元素打造自然科技的主题，在四个指廊候机区引用四季主题，运用不同的色彩和材质，使每个指廊各具特色，形成有温度的环境氛围；外遮阳构件提取了“鱼鳞”元素，室内天窗下方的采光带提取深圳市花“三角梅”元素，选用体现自然特点的主题构件；室内景观绿化蕴含“生态自然”等构想，打造“植物园里的机场”，在东西两侧各设置一个室外观景平台，为旅客提供一个全天候的室外观看飞机的空间，体验奔赴远方与回归家园的情景；针对办公人员，通过不同色彩的墙面喷涂来标识不同区位，丰富空间视野，以多种色彩打造人性化办公环境，改善员工的心理感受，增加对办公室的认同感。

2）动态情景交互

动态情景交互是近年来日渐常见的情景体验方式，已经进入越来越多类型的公共建筑中。在这种体验方式中，通过活动场所设计，运用创新科技的感知设施和展示技术、设置动态多样的场景条件，让人们主动参与进场景中，创造更具沉浸式的体验。在深圳机场卫星厅中运用了众多新型交互技术。通过设计时光隧道式捷运 + 未来感的捷运站，让旅客感受一种动态的科技穿越效果；通过鹏城天幕 + 双子星广场体验科技引领，结合 LED 大屏、灯带、影像营造兼具自然和未来感的建筑环境；东北侧平台设置飞行体验区、VR 体验、科技产品展示区，使旅客进行多种参与式的设施体验，营造温馨、舒适的场所氛围；针对不同年龄的旅客群体的需求，设置不同主题的儿童游乐区与多功能厅，提供针对性服务，增加卫星厅的动态体验感。

深圳宝安机场卫星厅鹏城天幕 + 双子星广场
Pengcheng Canopy + Gemini Plaza of Satellite Hall of Shenzhen Bao'an International Airport

2.6 HUMANISTIC INTERVENTION AND EXPERIENTIAL PERCEPTION

Humanity, as an important feature of regional architecture, embodies natural and social development of the place where the architecture is located and emphasizes humanized experience. Intervening and perceiving humanity means to identify the context that can inspire design starting with the analysis of the local environment, and to transform it into design content with cultural quality. The translation of cultural context can work with human activities to create multiple communication places, improve architectural quality via humanistic scenes, and inherit the "people-oriented" spirit of contemporary architecture.

2.6.1 Contextual Element of Image

As a concept closely associated with culture, CONTEXT points out the contractual relation between cultural elements and between culture and environment. Context has become a hot topic in architectural practice and research since Robert Venturi first introduced it into the field of modern architecture. The concept of context has been widely applied in the engineering practice of many fields, and become a basic component of the current human settlement discipline theory. Buildings in Lingnan with distinctive regional characteristics contain great historical and cultural value and simple, ingenious construction wisdom. They are the wellheads of contemporary architectural design in the Lingnan region. All-inclusive cultural heritages are valuable assets for architectural creation. Combining design practice with context characteristics can bring modern architecture and regional tradition together, and promote the inheritance and development of traditional culture.

1) Study on Context Memory

As global cultural exchange advances to a deeper level, the architectural styles of different regions tend to influence each other and become homogeneous. In spite of that, regional ethos need to be inherited to maintain cultural diversity. Contemporary Lingnan architecture in the process of globalization should combine local ethos and the characteristics of the times and maintain the representation of local language through the study of context memory. To design the GTC in front of T3, Guangzhou Baiyun International Airport, an in-depth study was conducted on the historical context of the city. The GTC features a building volume connecting the terminal in the middle and the six parking structures on the left and right sides, symbolizing the story of "one stalk with six ears" given to local people by an immortal in the legend of five rams. The interior design of the GTC continues the historical context of Lingnan hometown water to create a "new hometown water" landscape which joins the past with the present, and, with the hierarchical greening and cascade waterscape, presents a new city parlor with Lingnan charm.

2) Translation of Cultural Symbols

Translation in the very beginning was a linguistic act translating one language or character into another. Today, the conceptual connotation of translation has long been extended beyond the field of linguistics. However, the meaning of translation remains the same in essence as the conversion of one semantic system into another as per a certain rule, which is of great guiding significance to the expression of local architectural culture. The design of T2, Guangzhou Baiyun International Airport explores the contemporary translation of traditional Lingnan courtyard and architecture. In terms of space, it creates a Lingnan garden to offer passengers a new experience of contemporary Lingnan courtyards. In terms of form, it draws inspiration from the side-by-side roof features of traditional Lingnan architecture, and designs the top of the vertical green wall in the garden into a freely undulating broken line, inventing a new form of Lingnan architecture that is flexible and light. In terms of material, it simulates the distinctive texture of Lingnan architecture to develop three types of skins made of fair faced cultural bricks, U-shaped glass and fritted glass respectively, presenting a modern version of the building skin for traditional Lingnan architecture.

2.6.2 Popular Places for Communication

Places for communication exist widely in various public buildings. Toyo Ito believes that it's far more meaningful for public buildings to "gather a lot of people" than performing any specific function. Jan Gehl also proposes in his book Life between Buildings that places for communication play an important role in the vitality of public space. The creation of places for communication is not only conducive to the pluralistic development of architectural functions, but also greatly beneficial to a city's cultural vitality.

Focusing on the gathering and communication attributes of urban public spaces and enhancing the "human touch" of architectural places is of positive significance in today's design practice. The appearance of places is accompanied by the consideration of identity, emotional belonging, psychological security, etc. Therefore, the creation of places for communication should identify how people communicate with each other based on space and how to enhance the historical and cultural weight of space.

1) Injection of Vitality

The design of a transportation building may, through constant exploration of additional services beyond transportation, bestow more possibilities to the place, turn it from a transit place to a pleasant, vibrant destination that attracts numerous footfalls. The design of an urban industrial building, in addition to meeting the mandatory functional needs, can incorporate the building into the pluralistic urban environment to release multi-dimensional value in terms of publicness and create a promising urban vitality spot. The GTC in the new terminal area of Shenzhen Bao'an International Airport is a multi-modal transportation core of diverse functions. More than a traditional transportation center, it joins various retail and F&B functions and offers a variety of additional services. Lieqiao Substation, Guangzhou also goes beyond the concept of traditional version. It incorporates a science museum, exhibition hall and garden into the facility, and, together with the river view and cityscape nearby, makes the once cold NIMBY industrial substation part of the diversified urban spaces.

2) Humanized Design

Humanized design combined with the physiological features, psychological factors and behavioral habits of people can upgrade the quality of space and promote interaction between people and between people and buildings or cities, showing respect for and humanistic care to each individual. In T2, Guangzhou Baiyun International Airport, the passenger waiting hall is designed with baby care room, children's play area and cultural exhibition area; all counters and seats are designed in combination with ergonomics and barrier-free use demands, while all seats are equipped with a charging device on the armrest to offer considerate service to different groups of people. Lieqiao Substation, Guangzhou incorporates humanistic care into the industrial building, turning it from a "NIMBY facility" into a cultural leisure node of the city. Its modest, mellow shape perfectly harmonizes with the surrounding urban landmarks, while its rational, restrained modular facade reflects the due characteristics of industrial buildings. Its building skin made of perforated aluminum plates gives a clean and light visual effect, presenting a unique artistic effect of a "Moonlight Treasure Box" and adding a new great photogenic spot to the night scene of the Pearl River in Guangzhou.

2.6.3 Static Theme and Dynamic Interaction of Scenes

People's perception of space does not rely solely on vision. The introduction of atmosphere scene design into space composition can enrich people's perception of scenes. Peter Zumthor once mentioned, "In fact, all successful architecture seeks to convey people's comprehensive perception, which depends exactly on people's feelings and experience in space." In architectural design, architects should also pay attention to the comprehensive scene experience of people. Architectural scenes are beyond visual creation. They trigger various senses of people, such as weight, temperature, luster, surface roughness or smoothness, and even smell, which combine to form some kind of comprehensive perception. In public buildings, a scene theme and humanized design can marry rich experience with architectural scene, shape the overall impression of architectural scene perception, and enhance the humanistic appeal of architecture.

1) Static Scene Theme

While presenting static scene themes, the creation of multiple themes such as nature and humanity gives rise to an integrated architectural scene atmosphere. Shenzhen Airport Satellite Concourse employs a variety of elements to define the theme of nature & technology. The four pier waiting areas are themed on spring, summer, autumn and winter respectively using different colors and textures to foster a caring atmosphere. The external shading components resemble the "fish scale", while the daylighting belt under the interior skylight imitates the "triangle plum", the city flower of Shenzhen, both featuring theme components reflecting the characteristics of nature. The interior landscape greening reflects such ideas as "ecological nature" for the purpose of building an "airport in a botanical garden". On the east and west sides, two viewing platforms provide passengers with an amazing view into aircraft and an experience of the scenes where people either head to a faraway destination or return home. In the office area, different zones are marked with different colors of wall spray coating to enrich the spatial experience, better the employees' psychological feelings and enhance their recognition of the office space.

2) Dynamic Scene Interaction

Dynamic scene interaction is more and more commonly seen in recent years in a wide array of public building typologies as a way to experience scenes. Dynamic, diverse scene conditions created by innovative perceptual facilities and display technologies can enable a more immersive experience. Shenzhen Airport Satellite Concourse employs many new interactive technologies. The time tunnel APM and futuristic station offer passengers a dynamic and sci-tech experience. The Pengcheng Canopy and Gemini Plaza allow passengers to experience the leading technologies and the built environment, which is both natural and futuristic with large LED screen, light strip and imagery. The flight experience, VR experience and technology product exhibition on the northeast platform engage passengers with interactive amenities. Children's play areas and multi-purpose halls with different themes are also planned for people of different age groups to enhance the dynamic experience.

2.7 成本控制与价值驱动

建筑业作为国民经济的支柱行业，既受制于社会经济发展水平的高低，反之也影响着社会经济的发展。改革开放后，我国建筑业跟随时代蓬勃发展，并创造了巨大的经济效益。当今建筑行业已从高速发展向高质量发展逐渐过渡，建筑的价值不仅体现在其能带来的直接经济效益，还体现在建筑对行业、对社会作出的贡献，是社会、经济、艺术、科学多重价值的叠加。

2.7.1 全生命周期成本控制

全生命周期成本控制最早由英美工程界于20世纪80年代左右提出，通过多种分析手段，评估与决策出工程项目建设中及未来运营、维护的总成本最少的一种设计、建造、管理方式。建筑的成本主要包含建造与运营成本，对成本进行有效控制，可直接带来经济效益的提升。大型公共建筑因为投资较大，尤其需要拥有全生命周期控制成本的观念。

1）迎合定位、适度设计

迎合定位指项目应从发展战略的高度上思考设计，构建与项目战略相适应的建筑设计、建造、运营的全生命周期视角，设置合理的建设分段，通过适度设计实现成本的控制。以广州白云机场为例，T1航站楼的设计年旅客量为3,500万人次，在广州举办亚运会前，通过合理搭配与改造，实现了超6,000万人次的服务；T2航站楼预留与T1的连接楼与捷运系统，实现未来可持续发展，集中式国际国内安检厅、合并国内转国际与国际出发三关流程，实现枢纽资源的集约化利用；T3航站楼靠近现有城轨布置，通过构型的优化预留充分

广州白云机场 T1-T3 航站楼分期设计建设
Phased design and development of T1 & T2 &T3, Guangzhou Baiyun International Airport

的陆侧发展用地，未来可方便地扩建指廊，其 GTC 高效整合多种功能，进行分期建设，酒店入驻前后灵活利用建筑，释放出了宝贵的航站区空间资源。

2）节点完备、精细控制

大型公建往往具有复杂的节点体系，对其优化设计能有效减少材料使用。通过材料模数的统一、构造性能的提升、多专业一体化设计等全方位的精细控制，实现节点优化，并有效降低施工难度，从而保证建筑成本的控制。以广州白云机场 T2 航站楼为例，其在幕墙系统以及张拉膜系统的设计上，通过精细把控各专项设计完备度，确定统一模数，减少幕墙钢结构用量，将设备管线与幕墙二次结构进行一体化整合，方便现场施工，保证整体效果。

3）集约便捷、高效运维

集约便捷、高效运维是建筑使用需求的体现，在设计阶段应对流程进行优化把控，降低建筑运维成本。以广州白云机场 T2 航站楼为例，设计中对流程和运维进行了有益考虑。T2 航站楼采用集约的“指廊式 + 前列式”混合构型，站坪运行效率高；陆侧交通与空侧容量相匹配，各种交通方式无缝链接，高效便捷；旅客进入航站楼后，从办票到安检为单一方向设计，结合国际、国内安检的前置与并置，有效提高通关效率、提升旅客体验；国内采用混流流程，国内 / 国际采用可切换的混合机位，通过登机桥固定端实现功能衔接与切换，降低投资及运维难度。

2.7.2 示范效应价值驱动

建筑活动会带来一定的社会影响，良好的建筑实践可以发挥建筑的示范效应，促进社会价值的发展。通过经验的积累，不断优化行业实践，打造建筑精品，努力达到行业前沿，并引导行业标准的建立，有助于改善和推动社会的发展，驱动建筑多维价值的提升。

要想实现建筑实践示范效应，不仅需要探索先进技术，勇于创新，同时也要在设计、建造品质上严格把控，实现从理念到实践的高标准要求。城市主要机场航站楼通常定位较高，代表城市形象，体现出先进技术的集中使用，反映出建筑价值的驱动和引领作用。

1）技术创新促进行业进步

技术创新是推动行业良性发展的重要动力，通过技术创新，不仅有助于提升行业效率、优化人性体验，也能带来示范效应。以机场航站楼设计实践为例，不断实践新型设计理念，探索新技术的使用，提出高效与可持续发展的构型，优化交通核心效能指标和数据，实

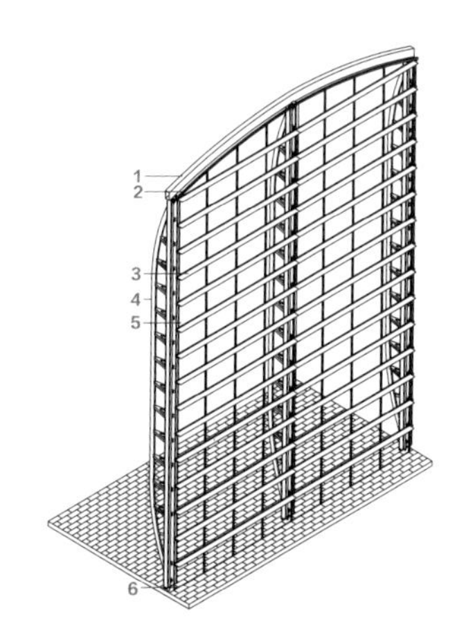

1. 幕墙顶部弧形箱型梁
Arc Box-Shaped Beam on Top of Curtain Wall
2. 弧形铝合金横梁
Arc Aluminum Alloy Beam
3. 铝合金横梁
Aluminum Alloy Beam
4. 钢桁架抗风柱
Steel Truss Wind-Resistant Column
5. 中空夹胶钢化彩釉玻璃
Insulated Laminated and Tempered Fritted Glass
6. U 型钢收边底槽
U-steel Bottom Cut

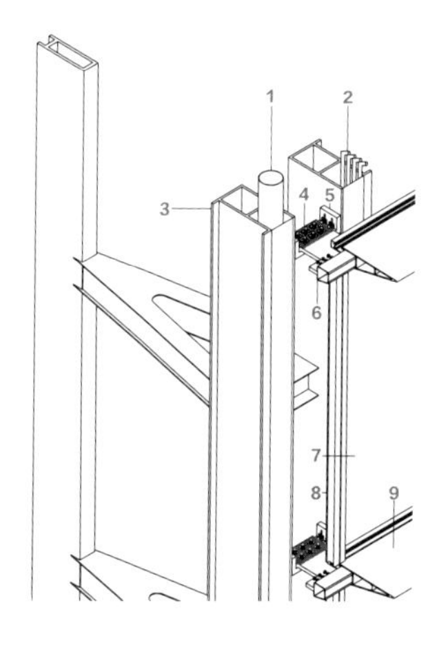

1. 落水管
Downpipe
2. 电气管线
Electrical Pipeline
3. 钢桁架抗风柱
Steel Truss Wind-Resistant Column
4. 铝型材转接件
Aluminum Profile U-steel
5. U 型钢转接件
U-steel Adaptor
6. 铝型材套管
Aluminum Profile Sleeve
7. 中空夹胶钢化彩釉玻璃
Insulated, Laminated and Tempered Gritted Glass
8. 竖向不锈钢拉杆外加铝盖板
Vertical SS Rod with Aluminum Capping
9. 铝合金横梁
Aluminum Alloy Beam

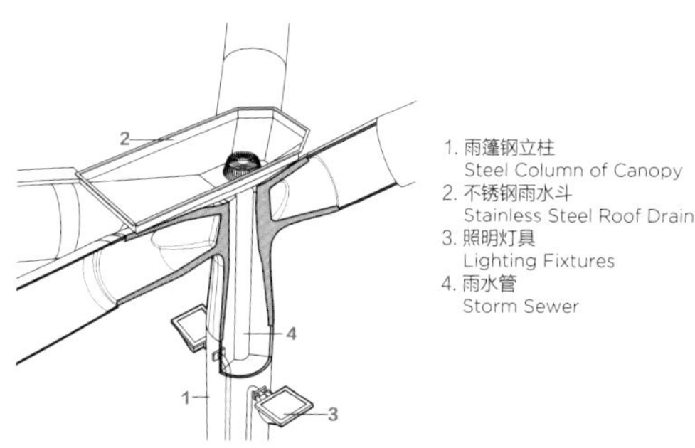

1. 雨篷钢立柱
Steel Column of Canopy
2. 不锈钢雨水斗
Stainless Steel Roof Drain
3. 照明灯具
Lighting Fixtures
4. 雨水管
Storm Sewer

广州白云机场 T2 航站楼幕墙系统，张拉膜系统单元节点示意
Curtain wall system & tensile membrane system unit nodes of T2, Guangzhou Baiyun International Airport

顺德德胜体育中心 4 个场馆集中同一地块有效节省用地
Four venues planned on one plot to save land used by Shunde Desheng Sports Center

现节约航站楼建设用地和规模，缩短航站楼施工周期的效果；提出“大安检区”概念，实现了旅客、行李、货物的综合联运，提升旅客服务体验、创造机场运营新价值点；气候适应性外围护结构设计填补相关空白，首次在航站楼应用被动防热节能围护构造技术，在华南地区强风频发、海边盐雾环境、降雨充沛的环境、气候特点下，实践并应用了航站楼屋面、幕墙等外维护结构的防台风设计关键技术。在多专业协同下，建筑技术得以不断发展，促进整个行业的技术进步。

2）品质把控引领价值提升

随着社会经济水平提升，人们对建筑品质的要求也越来越高，高标准的建筑品质把控可实现建筑的综合价值。公共建筑作为服务广大民众的建筑类型，应通过高品质的设计和建造，力求实现建筑的经济、社会、艺术、科学等多重价值。以广州白云机场 T1 航站楼、广州亚运馆、顺德德胜体育中心及广州猎桥变电站为例，在 T1 航站楼设计中，团队对机场类型建筑进行了全面系统的技术研究，通过结构、系统、工艺、运营的优化，使其成为面向世界的现代化门户；广州亚运馆是 2010 年亚运会的主场馆，复杂非线性三维曲面造型使其成为广州亚运会最为夺目的标志性体育场馆，具有很高的艺术价值；顺德德胜体育中心将场馆集中布置，节约的用地留给城市作为 TOD 开发，大大提升了用地价值；广州猎桥变电站在满足变电站的功能之外，还营造城市开放空间，打造为城市小客厅和电力展览教育基地，创造更多的社会价值。

2.7 COST CONTROL AND VALUE-DRIVEN ORIENTATION

As a mainstay of the national economy, the construction industry is constrained by yet can also affect the level of socio-economic development. After the reform and opening-up was initiated, China's construction industry has developed vigorously and created huge economic benefits. Today when the construction industry has gradually transited from high-speed development to high-quality development, the value of architecture is not only about the direct economic benefits it can bring but also the contribution to the industry and society. It is built on the values of society, economy, art and science.

2.7.1 Life-cycle Cost Control

The concept of life-cycle cost control was first proposed by the British and American engineering circles in the 1980s. It identifies the design, construction and management approach that can minimize the total cost during the construction, operation and maintenance of an engineering project through a variety of analysis methods. The cost of a building mainly includes construction and operation expenses, the effective control of which can directly increase economic benefits. In particular, large public buildings involving huge investment need to bring in the concept of life-cycle cost control in design.

1) Positioning-oriented and Moderate Design

Positioning-oriented design means to design a project from the height of development strategy. It requires to establish a life-cycle perspective of architectural design, construction and operation that is compatible with the project strategy, define reasonable construction segments, and achieve cost control through moderate design. For example, in Guangzhou Baiyun International Airport, T1 has a design capacity of 35 million passengers per year, but the terminal managed to serve over 60 million passengers before the 2010 Asian Games through reasonable functional combination and renovation; T2 reserves conditions for connecting buildings and an APM system to ensure sustainable development in the future; T3 is planned close to the existing urban rail transit and, with an optimized configuration, reserves sufficient land for landside development and pier expansion in the future. The GTC is built by phases to effectively integrate integrate various functions. The flexible use of the hotel building before and after the operation releases valuable spaces for aviation development.

2) Complete Nodes and Precise Control

Large public buildings often have complex node systems, so optimal design of nodes can effectively reduce the use of materials. Meanwhile, all-round precise control measures, such as unified material modulus, improved structural performance and integrated multi-disciplinary design, can optimize the nodes and effectively lower construction difficulty thus ensure the control of construction cost. For example, in the design of curtain wall system and tensioned membrane system of T2, Guangzhou Baiyun International Airport, precise control measures are taken to ensure the completeness of each specialty design, finalize the unified modules, and cut steel consumption of the curtain wall. Also, the MEP pipelines are integrated with the secondary structure of the curtain wall to realize easy on-site construction and desired overall effect.

3) Intensive, Convenient and Efficient Operation and Maintenance

Intensive, convenient and efficient operation and maintenance embodies the demands of the building users. The process should be optimized and controlled early at the design stage to cut building operation & maintenance cost. For example, the process and operation & maintenance of T2, Guangzhou Baiyun International Airport are fully considered in the design. The intensive "pier + linear" hybrid configuration of T2 ensures high apron operation efficiency. The landside traffic and airside capacity are matched, allowing seamless, efficient and convenient transfer between various transportation means. The single direction design of passenger handling process from ticket check-in to security check, coupled with the pre-positioning and juxtaposition of international and domestic security checks, effectively improve the clearance efficiency and passenger experience. To realize mixed flow process for domestic flights and switchable mixed stands for both domestic and international flights, T2 employs fixed end of boarding bridge for functional connection and switching, which cuts investment and facilitates operation and maintenance.

2.7.2 Demonstration Effect and Value-driven Orientation

Construction activities are bound to bring about some social impacts. Good architectural practice can give play to the demonstration effect of architecture and promote the development of social values. Accumulation of project experience, constant optimization of professional practices, creation of excellent architectural works, and establishment of industrial standards will all help drive the societal development and enhance the multi-dimensional value of architecture. Realizing the demonstration effect of architectural practices entails the exploration of advanced technology, the courage to bring forth new ideas, strict control over design and construction quality, and high-standard practice of concepts and ideas. As the name cards of cities, airport terminals are often highly positioned, reflecting massive use of leading technologies and showcasing the driving and leading role of architectural value.

1) Technology-driven Progress of Industry

Technological innovation is an important enabler for healthy industrial development. It can help improve industry efficiency, optimize humanized experience and bring about demonstration effects. In airport terminal design practice, we constantly practice new design concepts, explore the use of new technologies, and propose efficient, sustainable configurations, optimize key traffic efficiency indicators and data, so as to save construction land, reduce size of terminals, and shorten the construction period. We propose the concept of "large security check area" which helps, realize the integrated transportation of passengers, baggage and cargo, improve passenger service experience and create new value points for airport operation. Our design of climate adaptive building envelope design fills in the relevant gaps. We are the first to apply passive thermal protection and energy saving envelope technology to airport terminals. Given the climate of South China marked by frequent gale, seaside salt-spray environment and abundant rainfall, we apply key technologies of typhoon prevention to the design of terminal roof, curtain wall and other external envelope structures. It is the collaboration between multiple disciplines that sustains the evolution of our construction technologies and allows us to contribute to the technical advance of the entire industry.

2) Value Enhancement through Quality Control

As society advances and the economy booms, people are becoming increasingly demanding for architectural quality. Strict control of architectural quality can realize the comprehensive value of buildings. As a building typology serving the general public, public buildings should demonstrate the economic, social, artistic and scientific value of architecture through high-quality design and construction. Examples in this regard include T1, Guangzhou Baiyun International Airport, Guangzhou Asian Games Gymnasium, Foshan Desheng Sports Center and Guangzhou Lieqiao Substation. In our design of T1, our team conducted systematic technical research on airport buildings and made the terminal a modern gateway opening up to the outside world through the optimization of structure, system, processes and operation. Guangzhou Asian Games Gymnasium is the main venue of the 2010 Asian Games. Its complex, non-linear three-dimensional curved surface makes it the most eye-catching landmark sports venue for Guangzhou Asian Games. Foshan Desheng Sports Center centralizes four main venues on one side of the site and leaves the land on the other side to the city for TOD project, creating higher value of land use. In the design for Guangzhou Lieqiao Substation, spaces open to the city are created, including a small city parlor and an electric power exhibition and education base in addition to the substation function, enhancing the social value of the project.

2.8 前沿探索与创新实践

随着社会经济水平的发展与科学技术的进步，当代建筑理念与技术也不断创新。在多年的大型复杂公共建筑设计实践中，通过不断的探索，攻克众多核心关键技术，以点带面，形成全专业健全梯次的创新体系，不断培育新的创新点与增长点。展望未来，建筑必将进一步融入人类社会发展的进程，朝着更为人本化、生态化与智能化的方向前行。

2.8.1 探索创新实践

过去 20 余年，中国经历了世界上规模最大的建设进程，我们在与世界知名建筑师事务所的新理念、新技术的交流中，在大型复杂大跨度公共建筑领域完成了由吸收学习到自主创新的转换。加强对新一代信息技术、人工智能、智能制造、新材料、新能源等世界科技前沿性技术的研究，聚焦“未来城市”“绿色双碳”“防灾减灾”“智能建造”等创新技术领域，以大型复杂航空枢纽建筑科技进步为引领点，形成若干具有制高点作用的成套技术体系。

1）高效布局创新技术

高效布局创新技术在航空枢纽建筑和体育建筑等公共建筑类型中进行了探索。航空枢纽越来越强调自身布局的交通效率，通过综合多式交通联运、无感安检互认、高度复合融合枢纽营造等技术创新实现交通能级的跃迁，整合融入城市整体交通网络之中，航空枢纽也不再是城市交通的终点站，而是具有规划可持续发展的城市发展极；体育建筑通过创新性的布局，将相互独立的布局转变为一体的设计，并整合文化、休闲、商业等多种业态，使建筑成为城市中的活力客厅。

2）绿色建造创新技术

传统意义上的航空枢纽是能源消耗巨大的同义词。通过研发“规划设计、施工建造、运行维护”的建筑全过程绿色建筑创新性技术。

广州白云机场 T3 航站楼构型呈现
Configuration of T3, Guangzhou Baiyun International Airport

	小主楼+多卫星 Small main building + multiple satellite halls	小主楼+大卫星 Small main building + large satellite hall	大主楼+小卫星 Large main building + small satellite hall	中主楼+中卫星 Medium-sized main building + medium-sized satellite hall	单一航站楼 Singular terminal building
近远期规模匹配性 Compatibility of long-and near-term development scale	3	2	3	1	3
不停航施工难度 Difficulty of construction without air service suspension	2	1	2	1	2
一期APM必要性 Necessity of APM for Phase I	1	1	3	1	3
投资额与运营成本 Investment amount and operation cost	1	2	3	3	3
站坪运行效率 Operation efficiency of the apron	3	2	2	2	1
与其他航站区关系 Relevance to other terminal areas	3	3	2	3	3
是否侵入其他区 Whether to intrude into other areas	2	2	2	2	2
楼内步行距离 Walking distance in the building	3	2	2	3	1
航司分配灵活度 Airlines distribution flexibility	2	3	3	2	3
与交通中心关系 Relationship with GTC	2	2	2	2	2
总分 Total	22	20	24	20	23

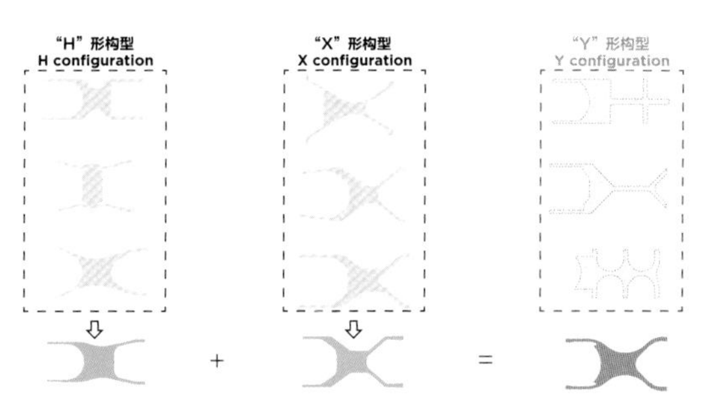

高效可持续航空构型方案比选过程
Comparison and selection process of efficient and sustainable airport configuration

在规划设计阶段，首次在航站楼建筑上采用被动防热、综合遮阳、大空间采光、光伏建筑一体化、近零能耗登机桥等绿色建筑关键技术，在设计阶段后期协同施工单位，首次采用航站楼绿色关键技术，在运行维护阶段，采用机电系统运维及后评估技术，为航空枢纽“绿色双碳”作出了探索。

3）应急保障创新技术

在应对极端气候和突发事件上，航空枢纽相关技术研究一直较为缺乏。规划设计阶段，创新岩溶地区基础设计、强风地区结构抗风、维护结构防台风、航站区海绵城市防涝减灾、平疫结合适应性改造等防灾安全保障关键技术，并率先提出应用多项防灾减灾设计技术，为实现“平安机场”目标提供了有力支撑。

4）全流程智慧应用创新技术

在人工智能智慧技术爆发的背景下，现有航空枢纽面向未来的智能智慧技术储备普遍不够。研发基于大型机场旅客、行李、货物、车辆全流程智慧出行保障，基于源网荷一环境一体化综合能源管控及设备节能控制等全流程智慧保障关键技术，应用了全生命周期全专业 BIM 协同精细化管理技术。

2.8.2 展望赋形未来

在现代建筑发展中，注重以人为中心设计、努力实现节能减排、积极应用科技杠杆等一系列变化，深刻影响了建筑的各个要素。在全球顶尖城市的中长期战略中，人本、生态、科技也成为未来城市聚焦的对象。我们期望通过前瞻性、引领性思考探索建筑、城市未来发展方向。

1）展望人类健康建筑

当前健康建筑的研究体现了对人本理念的关注。建筑设计迫切需要解决环境问题对人们健康的影响，对舒适环境的要求提升至了人类健康层面，健康舒适的空间品质、疫情新常态适应性建筑等成为可展望的发展趋势。我们以实践为基础，形成了《防疫建筑规划设计指南》等成果，积极探索适应岭南地区的未来健康社区、平疫结合运行的交通建筑、可灵活转换应急医疗功能的体育会展馆等新型健康建筑类型。

2）展望生态闭环建筑

恩佐·卡拉布赖斯提到，当下生态建筑方式并不能退化回到工业前的低能耗状态，在思考未来可持续发展建筑趋势时，应着眼于建筑资源循环、实现生态闭环的新思路。展望未来生态建筑，进一步挖掘如何

整合防台风技术的航站楼屋面系统
Terminal roof system integrating typhoon-resisting technologies

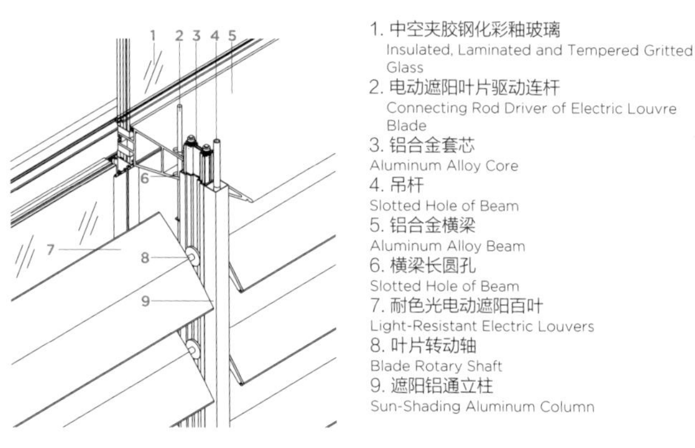

航站楼可调节电动遮阳百叶示意
Schematic diagram for adjustable electrical sun-shading lamella of the terminal building

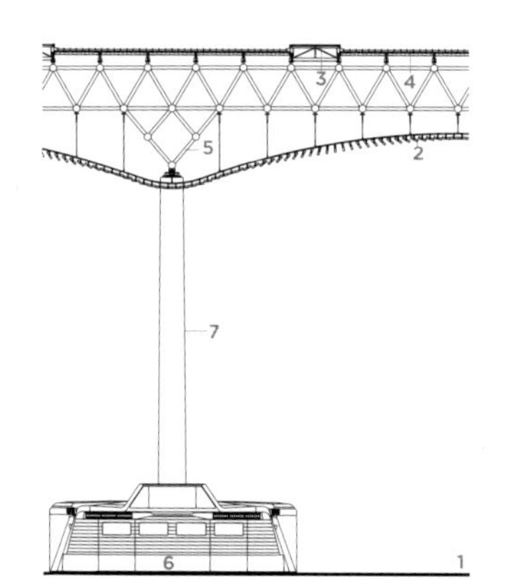

1. 办票大厅
Check-in Hall
2. 旋转天花叶片
Rotating Ceiling Blade
3. 采光天窗
Skylight
4. 直立锁边金属屋面
Vertical Edged Metal Roof
5. 带肋钢网架
Ribbed Steel Grid
6. 办票岛
Check-in Island
7. 钢管混凝土柱
Concrete Filled Steel Tubular Column

航站楼“长大带形天窗 + 渐变旋转式吊顶”采光设计
Terminal daylighting design of long and large strip skylight + gradient rotary ceiling

实现建筑空间、能源、材料改造再利用的可能。中国在不久的将来有望形成绿色能源为主的现代能源体系，促进建筑实践的低碳发展。在城市生态修复、多元能源协同供应、社区综合节能、“城市矿产”资源循环利用等方面通过探究，将构筑建筑全生命周期的闭环资源链条。

3）展望人工智能建筑

“移动城市”“漂浮城市”等探索曾体现了过去人们对智慧城市体系的展望，随着当今科技发展，智慧创新城市建筑的全领域技术图谱日益清晰，相关科技拥有着影响未来建筑发展的无限潜力。智慧城市运行、智慧交通管理、智慧社区管服、智慧市政设施互联互通技术正取得巨大进步；无人飞机、无人驾驶技术的成熟将给城市通行和建筑功能带来意想不到的变革；元宇宙、物联网等技术和理念的爆发，更助力城市建筑的科技升维，在未来，虚拟与现实的互动将更为紧密，现有建筑功能将被进一步解构。

曾经的科学设想正以前所未有的速度走向现实。面对科技前沿趋势，我们应该始终保持一种前瞻性和灵活性的态度，积极探索实践，以在科技变革的时代，助力未来建筑的发展。

2.8 FRONTIER EXPLORATION AND INNOVATION PRACTICE

Socio-economic development and technological advancement are the drivers behind the constant innovation of contemporary architectural concepts and technologies. After years of exploration and practice in designing large complex public buildings, we have resolved many key technology problems and established an innovation system that covers all disciplines and tiers, and continuously cultivated new points of innovation and growth. Looking ahead, architecture will be further integrated into the social development process of human beings and evolve toward a more humanistic, ecological and intelligent direction.

2.8.1 Innovation and Exploration Practices

Over the past 20 years, China has experienced the world's largest construction process. In the exchange of new ideas and technologies with world famous architects, we successfully transformed from learning to independent innovation in the field of large, complex, long-span public buildings. We strengthened research on the world's cutting-edge technologies such as next-generation IT, AI, intelligent manufacturing, new materials and new energy, and gave prominence to such innovative technology fields as "future city", "green building, carbon peak and carbon neutrality", "disaster prevention and mitigation", and "intelligent construction". Led by the technological advance of large complex aviation hub buildings, we have developed complete sets of technology systems of commanding heights.

1) Innovative Technology of Efficient Layout

We probed into the application of innovative efficient layout technology to aviation hubs, sports buildings and other types of public buildings. Aviation hubs nowadays tend to emphasize the transportation efficiency of their layout. Through technological innovation such as integrated multimodal transportation, automated security check and mutual recognition, as well as highly integrated hub construction, aviation hub buildings have been tremendously upgraded in transportation energy level and incorporated into the overall urban transportation network. They start to play the role of sustainable urban development poles instead of urban transportation terminals. As for sports buildings, an innovative layout can make their independent functions integrated and add culture, leisure and commerce to the mix to create a vibrant parlor in the city.

2) Innovative Technology of Green Construction

Traditional aviation hubs are synonymous with huge energy consumption. Through research and development, we have developed an innovative green building technology covering the entire process from planning and design to construction, operation and maintenance. In the planning and design stage, we pioneered in applying key green building technologies such as passive heat proof, comprehensive sunshade, large space lighting, building integrated photovoltaics, and nearly zero energy boarding bridge to a terminal building. We also took the initiative to adopt key green construction technologies for terminals in the construction stage and MEP system operation and post-evaluation technologies in the operation & maintenance stage.

3) Innovative Technology of Emergency Guarantee

The research on aviation hub technologies in response to extreme weather and emergencies has been relatively insufficient. We have creatively developed a number of key technologies for disaster prevention and security assurance in the planning and design stage of projects, including foundation design in karst areas, wind resistance of structures in areas with gales, the protection of building envelope against typhoon, sponge city waterlogging prevention and disaster reduction in terminal areas, and adaptive renovation serving both ordinary time and epidemic period. We are also the first to put forward and apply a number of disaster prevention and reduction construction technologies, providing strong support for achieving the goal of "safe airport".

4) Innovative Technology of Process-wide Intelligent Application

Faced with the rapid advance in AI technology, almost all aviation hubs today fall short in the reserve of future-oriented smart technology. For the purpose of guaranteeing process-wide intelligent mobility of passengers, baggage, goods and vehicles in large airports, and underpinned by the key technologies of process-wide smart guarantee, such as integrated energy management and control of power source, power grid, load and environment, and energy-saving control of equipment, we proposed life-cycle, all-discipline BIM collaborative fine management construction technology.

2.8.2 Vision into Future

Modern architecture values people-centered concept, energy efficiency and emission reduction, as well as active application of the scientific and technological level, profoundly influencing various elements of architecture. Top cities around the globe have also shifted their focus to people, ecology and technology in their medium and long-term development strategy. We expect to identify the future development direction of architecture and cities through forward-looking, leading thinking.

1) Vision into Healthy Buildings

The contemporary research on healthy buildings epitomizes people's attention to the humanistic philosophy. It is an urgent task for architectural design to address the impact of the environment on human health. With people's demand for a comfortable environment elevated to the level of human health, healthy, pleasant building spaces and buildings adaptive to the new normal of the pandemic have become the foreseeable future. On the basis of numerous practices, we have produced such achievements as the Guidelines for the Planning and Design of Epidemic Prevention Buildings, actively exploring new types of health buildings such as healthy future communities that adapt to the climate in Lingnan, transportation buildings that can operate during ordinary and epidemic periods, and sports exhibition halls that can flexibly switch to emergency medical functions.

2) Vision into Closed-Loop Eco-buildings

Architect Enzo Karabryce once mentioned that there's no way for current ecological buildings to return to their pre-industrial low-energy state. Our current reflection on the trend of sustainable building may be aimed at the new thought of realizing building resource circulation and ecological closed-loop. Our vision into future ecological buildings may further explore the possibility of transforming and recycling buildings, energy sources and materials. China expects to develop a modern energy system dominated by green energy sources such as hydropower, wind energy, solar energy and nuclear energy in the near future that can promote low-carbon building practice. Through exploration in urban ecological restoration, multi-energy coordinated supply, community comprehensive energy conservation, and "urban mineral" resource recycling, a closed-loop resource chain covering the whole life cycle of buildings will be built.

3) Vision into AI Buildings

Our previous exploration of "mobile city" and "floating city" reveals people's vision of a smart city system. Today, with the development of science and technology, the technical atlas of intelligent, innovative urban architecture in all fields is getting increasingly clear. Science and technology possesses infinite potential that can affect the development of future buildings. Great progress is being made in smart city operation, smart traffic management, smart community management service, and smart municipal facilities interconnection technologies. The maturity of unmanned aircraft and unmanned pilot technologies will bring unexpected changes to urban transportation and building functions. The explosion of technologies and concepts such as the metaverse and the Internet of Things will further boost the technological upgrading of urban buildings. In the future, the interaction between virtuality and reality will be closer, and the existing building functions will be further deconstructed.

The once imagined scientific ideas are turning into reality at an unprecedented speed. In the face of the cutting-edge technology, we should always maintain a forward-looking, flexible attitude and actively explore in practice to facilitate the development of future architecture in the era of scientific and technological change.

CHAPTER III

合和建筑观复合设计体系与闭环设计流程

INTEGRATED AND HARMONIOUS ARCHITECTURE: A COMPOSITE DESIGN SYSTEM AND CLOSED-LOOP DESIGN PROCESS

3.1 创新当代复杂条件下的设计路径

建筑设计实践条件随社会发展愈发复杂，在此情况下，更好地总结归纳，创新当代复杂条件下的设计路径显得尤为重要。建筑师应着重考虑多元要素的整合，充分整合自然、人文、历史、社会、使用者需求等各外部因素，评判其在设计中所起的积极作用或消极影响，规避风险并发掘潜在的有利因素；同时综合兼顾场地条件、功能、成本、技术等内部因素，统筹融合多学科，总结形成一系列科学设计方法。

3.1.1 多专业交叉融合

多专业交融是当代建筑设计的发展趋势，设计流程内各要素间、系统间、专业间互动互促，在大型复杂公共建筑领域内显得尤为重要，在协同中寻求最佳平衡点，形成有机高效联动。随着现代建筑发展的进程，建筑的复杂性与集约性趋势明显，建筑设计对于多专业的有效协同与融合度要求与日俱增，大型航站楼建筑设计是多专业融合的典例，建筑师不仅需要深入理解航站楼各方面的需求，还需要跟其他专业顾问密切配合，融合众多专业及复杂的技术体系。

以大型复杂航空枢纽为例，其包含规划设计、市政工程设计、建筑设计、结构设计、机电设计、民航工程设计等专业大项，涉及机电及民航、行李系统等专业的各分项系统设计更加细分庞杂，叠加建筑全生命周期维度后，还需纳入建筑策划及后评估、建设施工、运维管理等环节创新；相互交流会接触到不同专业的观点，例如建筑师想要创造出令人满意的艺术空间，结构更关注的是其安全性与稳定性，商业策划则想要达到其收益最大化，需善于运用这个时代所拥有的技术，从中寻求建筑整体的最佳平衡点。以深圳宝安机场卫星厅为例，其强调室内设计与建筑设计一体化，注重与结构设计紧密结合。通过建筑、结构、装修、幕墙的一体化协同，提炼出深圳市花三角梅和深圳机场 LOGO 的三角形元素，成为贯穿室内设计的三角形母题，天花设置三角形单元结构构件，天窗下的滤光器以三菱锥的形式，与旅客的流程充分结合；幕墙标准单元的外遮阳构件采用“L”形的类直角三角元素，贯穿整个表皮设计。通过二维协同设计与三维共享建模协同推进，使多专业有机融合，呈现出整体实践效果。

3.1.2 全专业多层次的绿色设计

随着时代发展，绿色设计已成为建筑设计中的重要环节。通过搭建全专业的绿色设计平台，多层次绿色体系建构，使建筑在整个生命周期中最大限度地提升绿色水平，改善建筑能耗，提升经济、社会效益。

1）全专业绿色平台搭建

绿色设计并非游离于建筑之外或是强加于建筑的独立系统，而是一种潜在的设计意识与设计原则，涉及所有专业的设计内容，融合在建筑的每一个系统与细节当中，不可分离。在广州白云机场 T2 航站楼的绿建设计中，通过搭建全专业绿色平台，将绿建设计融于全专业当中，形成系统性、整体性的绿建设计体系。全专业系统整体的绿建设计主要包含两个层次的内容：一是专业内部各系统的分解与整合，二是各专业之间各系统的协调、组织与整合。以全专业技术系统整体性设计的理念进行绿色设计，使绿色设计成为一种自觉的设计原则和设计行为，让其生于建筑，而又长于建筑。经过绿色设计平台的努力，绿色建筑已被打造为 T2 航站楼的一张亮丽名片，获得了良好的经济效益与社会效益。

2）多层次绿色体系建构

绿色设计拥有多层次的广泛内涵，与当地的自然条件息息相关。基于岭南地区大型复杂航空枢纽的设计实践，提出了一套具有地域适应性的绿色设计体系，包含高性能被动防热节能技术体系、室内外综合遮阳技术体系、大空间采光照明体系、光伏建筑一体化体系、近零能耗登机桥体系等绿色设计体系，解决航空枢纽建筑绿色节能先进性与舒适性问题，实现航空枢纽可再生能源高比例应用。

3.1.3 全生命周期与全专业 BIM 协同

大型公建涉及复杂设计因素的平衡与协同，需要强有力的设计平台和工具来推动。回顾历史，从图板绘图到 CAD 是一次飞跃，从 CAD 到 BIM 正处于第二次飞跃。以正向设计 BIM 模型、协同深化 BIM 模型、可计量管理 BIM 模型为平台，大型公建设计可进一步实现基于 BIM 的创新应用设计，有效开展多方设计协同、设计管理与进度投资管控。

1）基于大型复杂建筑的 BIM 创新应用设计

复杂建筑往往体量巨大、功能需求多样，传统的二维设计手段已无法满足其复杂设计需求。通过建立正向 BIM 模型，形成全方位进阶的技术手段。结合 Rhino 在复杂形体与空间的处理能力，利用

Rhino 导出复杂模型至 REVIT 并赋值转化成为信息模型，满足深化及算量要求，形成高效、高质的数字化设计模式；引入多种可视化分析手段，助力建筑空间方案综合论证与优化；通过枢纽交通运行仿真模拟，对交通环境进行分析与优化；针对复杂的构件构造，建构满足加工级精度的 BIM 信息模型。

2）基于多方参与的 BIM 协同工作与设计管理
大型项目涉及多种复杂的建筑功能，设计标准差异性大，同时项目管理方多，做好设计协同和管理协同是项目工作的难点，通过建立可供协同的深化 BIM 模型，做到设计标准协同、操作平台协同以及数据协同，做到设计和管理多方协同。

3）基于全流程的 BIM 项目进度与投资管控
BIM 技术应用为项目进度与投资管控提供了一种新思路和新工具，构建可满足计量管理的 BIM 模型成为必要手段。可实现工程量实时统计动态算量、工程现场进度与投资有效管控。

3.2 复杂大跨建筑设计数据分析体系

建筑设计中，通过不同的分析手段，形成建筑设计可量化、成体系的数据体系。其中，空间句法是一种较为成熟的定量研究方法，随着其理论不断修正和发展，运用范围也不断扩大，逐步从城市规划领域的运用拓展到建筑空间领域；在建筑环境的分析中，通过多要素、多方位的分析，指导设计理性决策；在大型公共交通建筑中，基于仿真模拟的流程分析也为设计提供了理性数据基础。

3.2.1 基于可量化的建筑空间分析

随着大数据时代到来及各种数据处理技术的成熟，以数据为主导的研究方式逐渐被引入建筑设计中，为建筑空间的定量研究提供技术基础。其中基于空间句法理论的空间可视性分析和基于眼动追踪技术的空间展陈方式分析为大型公共建筑空间设计提供了定量研究方法。

1）空间可视性分析
空间句法在如今已形成一套成熟完整的理论体系和专门的空间分析软件技术。其理论通过计算机软件模拟分析，量化空间与空间之间的可

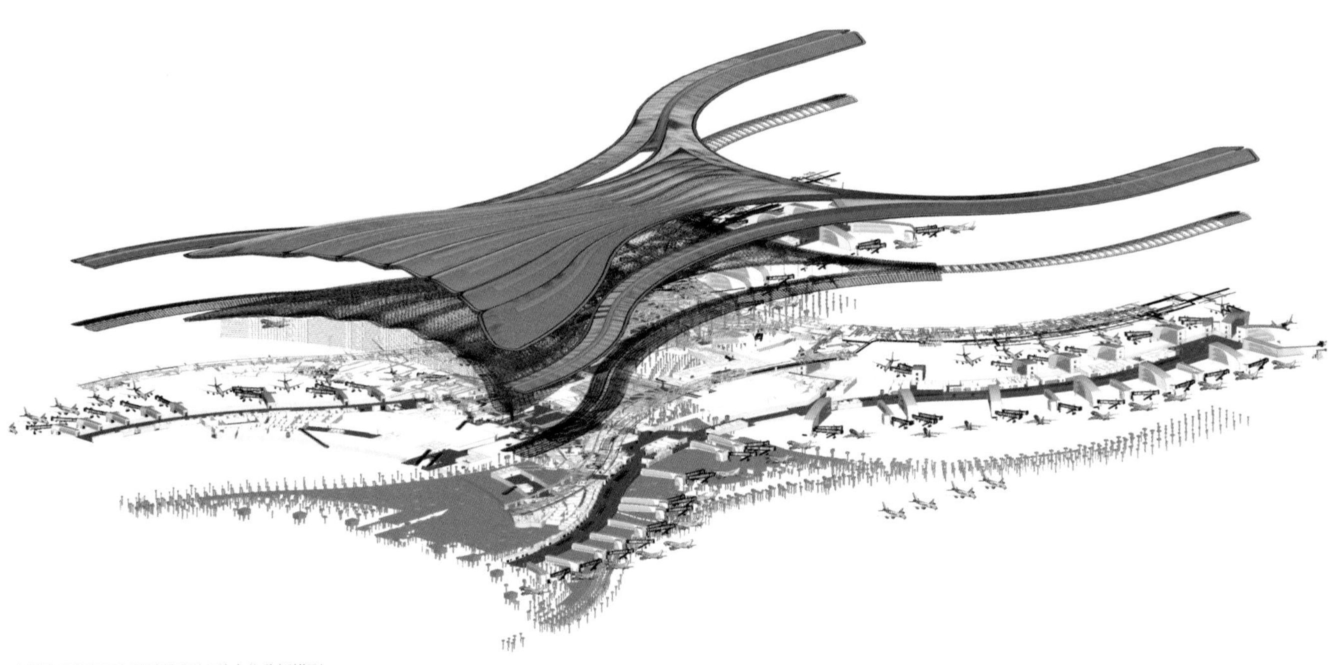

广州白云机场 T3 航站楼 BIM 数字化分析模型
BIM digital analysis model of T3, Guangzhou Baiyun International Airport

达性与可视性。在广州白云机场 T2 航站楼内部空间可视性分析中，通过对国内混流区的集中商业空间视线连接度进行模拟，显示该区域处于连接度较低的位置，商业区中庭周边空间的视线连接度差异较大。结果说明了旅客在混流区的集中商业空间内容易发生滞留、问询、折返等现象。另外，集中商业区内的柱子对空间可视性的影响较大。

2）空间展陈方式分析

通过眼动追踪技术可以将被试的感知及刺激材料视觉吸引力进行客观量化，分析人在建筑中不同空间的注意力分配程度，为建筑空间的展陈方式设计提供数据支撑。通过选取建筑中的典型空间的展陈方式，采用眼动追踪技术，分析在不同空间形态下，不同展陈方式的视觉吸引力程度以及人对不同空间的观察认知模式有何规律。以机场空间为例，空间引导性要素、展示物、透明性的空间界面均容易引起人的视觉关注，因此在机场空间设计中将这些内容进行设置，对展示效果提升及空间设计具有积极作用。

3.2.2 基于立体多维的建筑环境分析

建筑环境由多种要素共同组成，对建筑环境的分析需要从多维度切入，包括多要素、多方位等综合分析。

1）多要素综合分析

声、光、热是建筑环境的重要组成要素，大型公共建筑由于具有庞大的建筑规模和复杂的内部设计，平衡以上要素的难度更高。在广州白云机场 T2 航站楼项目中，通过 CFD 流体力学分析软件对室外风环境进行模拟，优化建筑平面设计和玻璃幕墙可开启位置及面积；通过动态能耗模拟软件和照明能耗模拟软件对建筑全年空调采暖能耗和照明能耗进行分析，指导暖通与照明合理设计。结合模拟分析结果，在进深较大的空间上方屋面设置均匀采光天窗，为室内提供良好的自然照明；结合室内视线效果与室外眩光影响，选用可见光反射率适宜的幕墙玻璃，减少光污染；通过对场地噪声进行模拟分析和场地绿化隔声带设计，改善建筑场地室外声环境。

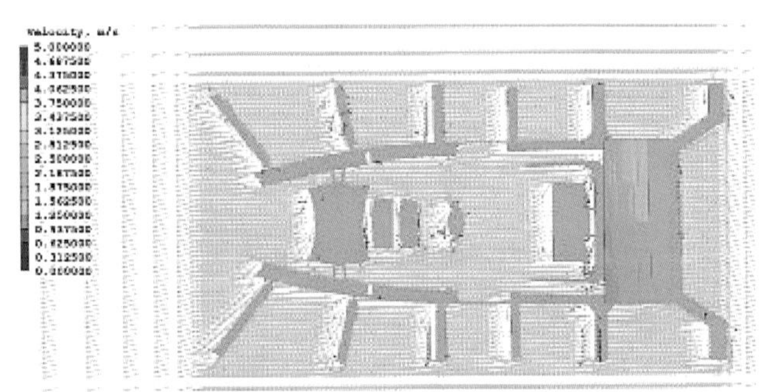

冬季室外1.5米高风速分布

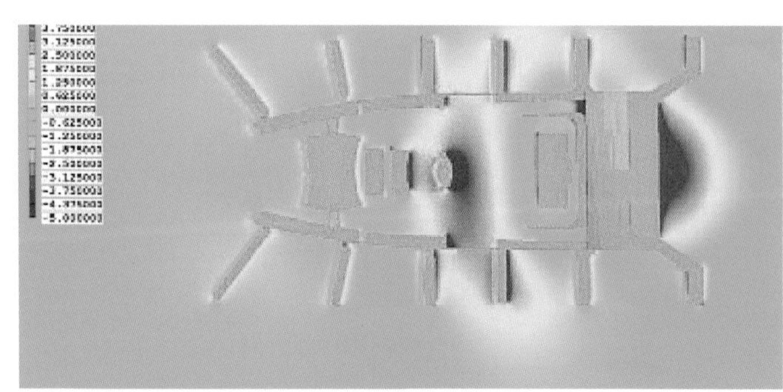

冬季室外1.5米高风压分布

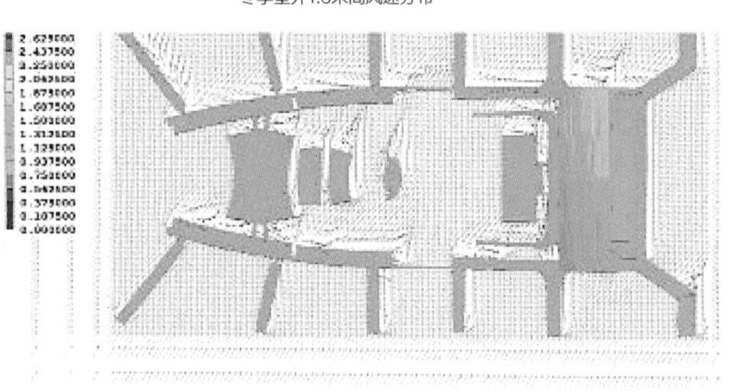

夏季室外1.5米高风速分布

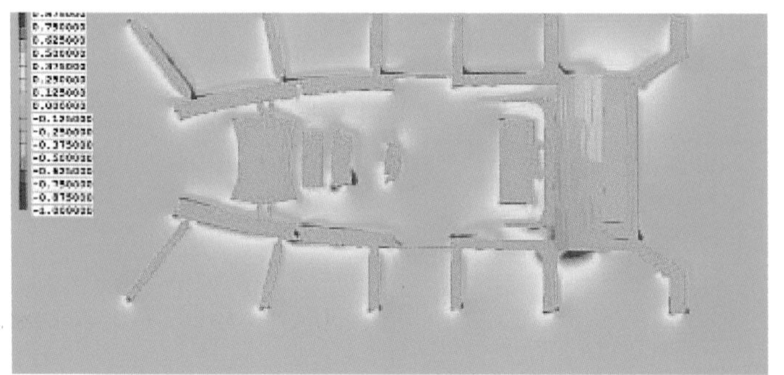

夏季室外1.5米高风压分布

广州白云机场 T2 航站楼室外热环境分析
Outdoor thermal environment analysis of T2, Guangzhou Baiyun International Airport

2）全方位立体分析
对建筑环境中某一要素采取多个角度的分析方式，有助于更加理性地指导设计决策。以珠海横琴保利中心为例，为打造可呼吸的绿色办公场所，通过软件模拟计算，构建了不同方位的“风道”，建筑中部设有露天天井，一层、二层均设有架空层。大楼通风效果主要关注点为自建筑底部风口经中庭后至建筑顶部能否形成完整的通风路径以及中庭内的风速分布情况，于是从多个方位研究分析了建筑室外风压及流场分布、主体建筑水平风道内风速分布和主体建筑竖向风道内风速分布，最终形成最利于大楼通风的建筑天井和架空层尺寸。

3.2.3 基于仿真模拟的流程分析

当代公共建筑的规模与复杂程度不断加大，各要素的仿真模拟分析能为建筑设计建立科学合理的评价体系。对于机场航站楼，应用仿真模拟有利于项目合理规划建设、完善服务水平、有效控制投资，实现从航站楼构型、陆侧交通（含 GTC）流线、航站楼内部流程的全方位仿真模拟分析，有效促进机场航站楼设计体系的完善。

1）航站楼构型仿真模拟
确定构型是机场航站楼设计的首要步骤，适宜的航站楼构型可以匹配机场发展定位和等级规模，满足旅客增长需要，发挥机场空陆侧交通效能，节约土地资源，实现绿色低碳与可持续发展。在广州白云机场 T3 航站楼项目设计实践中，将各个核心效能指标量化为具体可评价数据，与国内外同等规模机场进行横向大数据比选分析，实现因素综合最优。

2）陆侧交通（含 GTC）流线仿真模拟
陆侧交通是机场航站楼运行效率的重要保障，通过借助计算机模拟技术，高效组织各类交通流线，科学合理进行资源分配。以揭阳潮汕机场为例，通过基础数据及预测数据的交叉模拟，推导航站楼出发与到达车道边所需长度，在交通中心相应位置合理布置垂直交通设施，实现人流与车流的高效疏通与转换。

3）航站楼内部流程仿真模拟
机场航站楼是高度复合的交通建筑，高效的旅客流程与舒适的空间体验是航站楼服务标准的核心。在广州白云机场 T2 航站楼项目设计实践中，通过仿真模拟技术，助力打造五星级行业标准，利用 ARCPort 软件工具对客流密度、排队时间、步行距离等进行旅客全流程的仿真模拟，指导功能与设施设备分布，解决资源合理均衡分配。

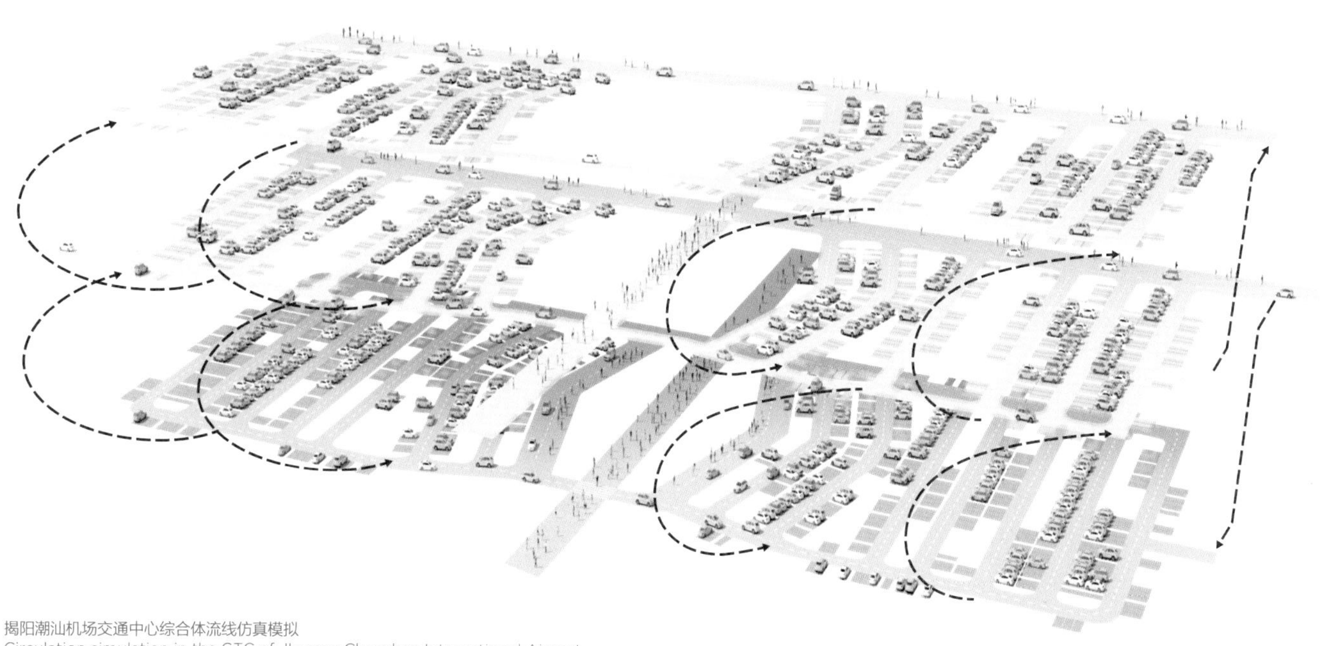

揭阳潮汕机场交通中心综合体流线仿真模拟
Circulation simulation in the GTC of Jieyang Chaoshan International Airport

3.3 建立互馈闭环的全周期全要素设计流程

3.3.1 策划及设计

建筑全周期闭环流程从开始就需要展开系统性、科学性、完整性的策划及设计。在广州白云机场 T2 的建设过程中，前期对机场发展定位、土地资源利用等上位规划、机场综合交通体系规划、地域适应性及场地适应性研究、航站楼规模功能研究、航站楼设计及运维系统建议、“四型机场”规划等进行系统性策划研究；同时对 T2 航站楼建筑设计实践进行分析和归纳总结，如各专业闭环设计的分析，以制定适宜的机场航站楼设计框架，形成完善的地区机场航站楼建筑设计的技术体系和方法。

3.3.2 施工协调

为保证建筑呈现最佳的设计效果，完成设计阶段后，还应全面对接协调现场施工。现场服务中，不仅对设计进行交底，还需解决施工中发现的与设计相关的现场问题，尤其在复杂的大型公建项目中，显得尤为重要。在现场协调过程中，不仅应与业主保持密切沟通，也应与施工单位多边配合。以广州亚运馆、汕头亚青会场馆为例，项目造型复杂，系统繁多，涉及钢结构系统、金属屋面板系统、幕墙系统、室内装饰设计、复杂曲面异型空间等。其三维的非线性曲面设计，相对而言，比常规的平面二维设计更加难于进行施工控制，从多种不规则曲面加工的平顺交接、精确定位，到核对监控、后期工艺等各环节均需全面考虑，确保现场施工完成度。

3.3.3 运维及后评估

运维及后评估是形成闭环设计流程的关键步骤。以广州白云机场 T2 航站楼为例，构建绿色低碳运维及后评估技术，通过一体化管理平台确保机电设备（系统）的设计、生产、安装、调试、验收及使用维护全生命周期信息的一致性、全面性与准确性，确保项目运维期间的操作规程、应急预案完善且有效实施。应用信息化手段进行物业管理，根据运行检测数据，进行设备系统的运行优化，并实现机电系统信息共享与资源整合，提高工作效率，提升运维效能。通过绿色低碳运维及后评估技术，极大地节约了资源，降低了航站楼建筑的运营和维护成本，打造全生命周期绿色低碳航站楼建筑。

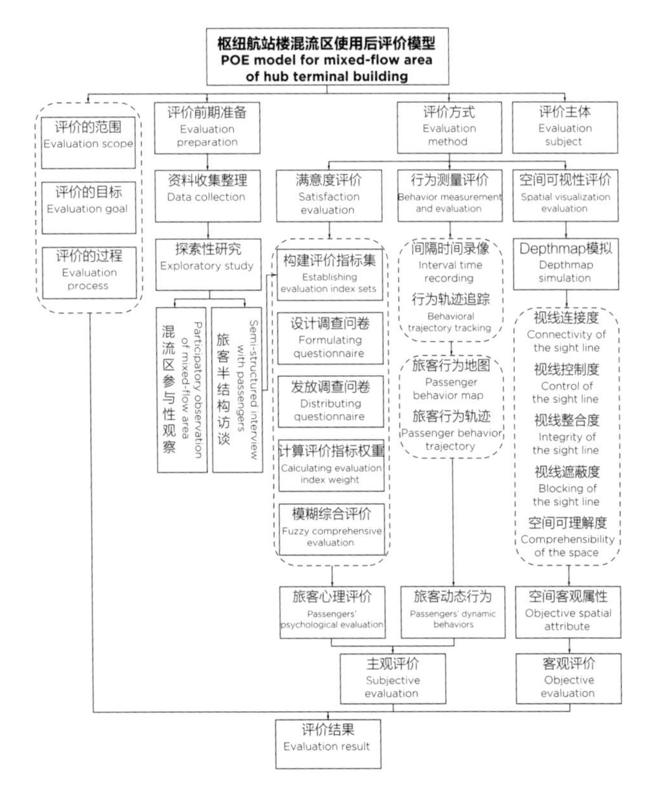

航站楼混流区使用后评价模型
POE model for mixed-flow area of terminal building

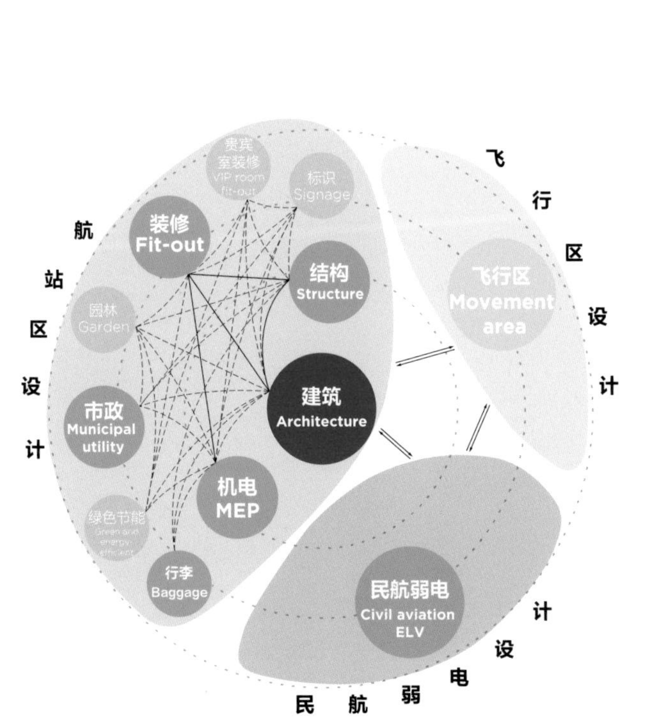

航站楼全专业互馈闭环设计模型
POE model for mixed-flow area of hub terminal building

3.1 DESIGN PATH INNOVATION UNDER COMPLEX CURRENT CONDITIONS

Architectural practice faces increasingly complex conditions along with the development of society. Against such background, it is particularly important to better summarize and innovate the design path today. Architects should focus on the integration of multiple elements, which means fully considering various external factors including nature, humanities, history, society, user needs and taking into account internal factors such as site conditions, functions, costs, and technologies. It is a multi-disciplinary undertaking that requires the summary of a series of scientific design approaches.

3.1.1 Interdisciplinary Integration

Interdisciplinary integration is a trend of contemporary architectural design, and the interaction and mutual promotion between the factors, systems and disciplines of the design process are more than important, especially in respect of complex public buildings, which requires the best balance of collaboration and an organic and efficient collaborative response mechanism. With the development of modern architecture, buildings tend to have more complex and intensive functions, while the requirement for effective synergy and integration of multiple disciplines is increasing in architectural design. Large airport terminals are a classic example of multi-disciplinary integration and architects not only have to develop a deep understanding of all aspects of the terminal, but also work closely with consultants of other disciplines and incorporate multiple disciplines and complicated technical systems into the design.

For example, large-scale complex airport hubs often involve major disciplines such as planning, municipal engineering, architecture, structure, M&E, civil aviation engineering, etc., as well as more detailed and complicated specialty design, such as M&E, civil aviation, and baggage system etc. Once the whole building lifecycle is added as a new perspective, other innovative links should also be considered, such as architectural planning and post-evaluation, construction, operation, maintenance and management, etc. Interdisciplinary exchange can present different professional views, while modern technology should be fully leveraged to strike the overall optimal balance for the building. For example, Shenzhen Satellite Hall emphasizes the integration of interior and architectural design, as well as the close combination with structural design. Through integrated synergy of architecture, structure, decoration and curtain wall, the triangular elements, which are extracted from triangular plum, Shenzhen's city flower, and the logo of Shenzhen Airport, are turned into motif throughout the interior design, the triangular structural components in the ceiling, and the light filter in triangular pyramid under the skylight to fully combine with the passenger process. The exterior shading of the typical curtain wall units also adopts an L-shaped component resembling a right angle throughout the building skin design. The collaborative 2D design and 3D shared modeling enable organic integration of different disciplines and expression of overall project effect.

3.1.2 Multi-Disciplinary and Multi-dimensional Green Design

With the development of the times, green design has become an important part of architectural design. The establishment of a multi-disciplinary green design platform and multi-dimensional green systems enables buildings to maximize their green building level during the entire life cycle, reduce energy consumption and enhance economic and social benefits.

1)Multi-disciplinary Green Design Platform

Green design is an underlying design awareness and design principle that involves contents of all disciplines and are integrated into and inseparable from each and every detail and system of a building. It is necessary to carry out green design with a holistic and multi-disciplinary design concept and make green design a conscious design principle and behavior. Green design should be born out of and grow together with buildings and help achieve economic and social benefits.

2)Multi-dimensional Green System

Green design has multi-dimensional and profound connotations and is closely related to local natural conditions. Based on the design practice of large-scale and complex aviation hubs in Lingnan region, a series of green design systems with regional adaptability are proposed, including high-performance passive heat-protection and energy-saving technology system, interior and exterior comprehensive shading technology system, large-space lighting system, photovoltaic building integration system, near-zero energy consumption boarding bridge system. They provide the latest and comfortable green and energy-saving solution to aviation hub buildings and realize high proportion of renewable energy used in aviation hubs.

3.1.3 Life-cycle and Multi-disciplinary BIM Coordination

Large-scale public buildings involve the balance and synergy of complex design factors, which requires a powerful design platform and tools. With BIM of forward design, collaborative design development and measurable management as platforms, BIM-based innovations and applications for large-scale public buildings help improve the efficiency of multi-party design collaboration, design management and cost/schedule control.

1)BIM-based Design Innovations and Applications for Large-scale Complex Buildings

Complex buildings are often huge in volume and diverse in function. Traditional two-dimensional design methods can no longer meet such complicated design needs. With a well-established forward design BIM, technical means can be improved in an all around way, including BIM-based space feasibility study and optimization, BIM analysis and optimization in traffic environment, and BIM design optimization under construction conditions.

2)BIM Coordination and Design Management Based on Multi-party Participation

Large projects are often undertaken by consortium of design teams, while design/management coordination remains a big challenge for the project due to complicated building functions, different design standards of design teams, and multiple project management teams. In this case, a deepened BIM for coordination can be established as a critical solution, and cover BIM coordination in design standards, platform and data.

3)Full-process-based BIM Project Schedule and Cost Control

The application of BIM technology provides a new idea and a new tool for project schedule and cost control. It is necessary to establish measurable BIM models to realize real-time statistics of dynamic bill of quantities, and effectively control the site progress and cost.

3.2 DESIGN DATA ANALYSIS SYSTEM FOR COMPLEX LARGE-SPAN BUILDINGS

In architectural design, it is necessary to build a quantitative and systematic data system through different analytical means. Among them, spatial syntax is a relatively mature quantitative research method, with the continuous revision and development of its theory, it is applied more and more widely from urban planning to architectural space. In the analysis of the building environment, total-factor and multi-dimensional analyses shall be carried out to facilitate the rational decision-making process of design. In large-scale public transportation buildings, simulation-based process analysis provides data foundation for design.

3.2.1 Building Space Analysis Based on Quantitative Standards

In recent years, with the coming of big data age and the maturity of various data processing technologies, data-focused research methodology has been gradually introduced into architectural design, providing a technical basis for quantitative research of architectural space. Among them, spatial visibility analysis based on spatial syntactic theory and spatial layout analysis based on eye tracking technology also provide quantitative research methods for spatial design of large public buildings.

1)Spatial Visibility Analysis

Space syntax now has been developed into a sound theory with specialized spatial analysis software. The theory can be analyzed through computer software simulation to quantify the accessibility and visibility between spaces. In the analysis of the interior space visibility of T2, Guangzhou Baiyun International Airport, the simulation of visual connectivity in centralized commercial space of domestic mixed flow area shows the area in low connectivity, and visual connectivity of spaces around commercial atrium vary greatly, indicating that passengers are prone to stay, inquire and turn back in this area. In addition, columns in the centralized commercial area also significantly compromise spatial visibility.

2)Spatial Layout Analysis

Through eye tracking technology, the perception of viewers and the visual attractiveness of stimulus materials can be quantified and the attention distribution on different spaces of the building can be analyzed, which provides data preparations for layout design of architectural space. By selecting different layouts of typical spaces in a building, eye tracking technology can be used to analyze their different visual attractiveness and draw a conclusion how people observe and perceive different spaces. Given the example of airport space, guiding spatial factors, display objects and transparent spatial interfaces are all visually attractive, and therefore, should be included in airport spatial design to positively improve the display effect and spatial design.

3.2.2 Multi-dimensional Building Environment Analysis

Building environment is composed of a variety of factors, and it is necessary to analyze it in a multi-factor, multi-dimensional and comprehensive way.

1) Multi-factor Comprehensive Analysis

Sound, light and thermal factors are important components of building environment and it is difficult to balance these factors in large public buildings. For T2, Guangzhou Baiyun international Airport, the outdoor wind environment is simulated by CFD, a fluid mechanics analysis software, to optimize the building plan and glass curtain wall; dynamic energy consumption simulation software is used to help the HVAC and lighting design, and evenly placed skylights are adopted to ensure daylighting, save electricity and reduce light pollution; moreover, simulation analysis on site noise and design of sound insulation median strips are conducted to improve the outdoor acoustic environment.

2) Multi-dimensional Analysis

Taking a multi-angle analysis on a factor in the building environment helps more rational design decision-makings. For example, Zhuhai Poly Center employs Software simulations to build "wind passages" on various directions based on analyses on the distribution of outdoor wind pressure and flow field and the distributions of wind speed in the horizontal and vertical stacks of the main building.

3.2.3 Simulation-based Process Analysis

With the development and popularization of computer technology, simulation technology has been used in many industries, using digital models for analysis and testing. For example, the airport terminal design practice, based on long-term study and practice, has realized all-round simulation and analysis in terms of terminal configuration, landside traffic (including GTC) circulation and processes inside the terminal to effectively improve the design system for airport terminals.

1)Terminal Configuration Simulation

An efficient configuration needs to fully reflect key performance indicators such as efficiency, capacity, humanization, and carbon emission under the premise of ensuring safe operation. In the design practice of T3, Guangzhou Baiyun International Airport, each key performance index is quantified into specific evaluable data, and computer simulation technology is applied to achieve a comprehensive optimal solution.

2)Landside Traffic (including GTC) Circulation Simulation

Landside traffic is an important guarantee for the operational efficiency of airport terminals, which requires efficient landside design. For example, in the design of Jieyang Chaoshan International Airport, based on cross-simulation of basic and projected data, the lengths required for departure and arrival curbsides of terminal building are calculated and a double-deck curbside is planned to satisfy the simulation indicators. Analysis-based design optimization helps realize efficient distribution and transfer of pedestrian and vehicular flows.

3)Simulation of Processes inside Terminal

Efficient passenger processes and comfortable space experience are essential to airport terminal service standards. With simulation technologies, Guangzhou Baiyun International Airport T2 terminal is able to meet the five-star standard of the industry at lower energy consumption and cost. With ARCPort software, it simulates all passenger processes to guide the distribution of functions and facilities and achieve the reasonable and balanced allocation of resources.

3.3 A MUTUAL-FED, CLOSED-LOOP, LIFE-CYCLE AND TOTAL-FACTOR DESIGN PROCESS

Architectural design involves many disciplines and contractors thus requires real-time and efficient collaboration in the whole development process. Architects are responsible for the life-cycle and total-factor control of projects and are specially required to have the ability to implement the whole process from project planning, design, construction coordination, operation and maintenance to post-evaluation.

3.3.1 Planning and Design

The life-cycle closed-loop process starts from systematic, scientific and complete planning and design. In the development of T2, Guangzhou Baiyun International Airport, we carried out systematic planning and research on airport development orientation, the use of land resources, airport comprehensive transportation system and regional adaptability in the initial stage and, based on the analyses of closed-loop design for each specialty, summarized the architectural design practice of T2 to establish an appropriate airport terminal design framework.

3.3.2 Construction Coordination

In order to ensure the best design effect, architects should also coordinate with the construction contractor on the site after the design is completed. On-site coordination services shall include both the clarification of design issues and the resolution of design problems encountered in construction. Taking Guangzhou Asian Games Gymnasium and Shantou Asian Youth Games Stadium as examples, they both involve complicated shapes and various systems, including steel structure system, roof system, curtain wall system and interior decoration design. Their three-dimensional nonlinear surface design is much more difficult to control as compared with conventional two-dimensional design. The architects have to consider all the aspects in order to ensure the perfect completion of construction, including smooth joint and precise positions of irregular surfaces, on-site check and post-process techniques.

3.3.3 Operation, Maintenance and Post-evaluation

Operation, maintenance and post-evaluation are critical steps in a closed-loop design process. For example, a green and low-carbon operation and maintenance and post-evaluation technical system is established for T2, Guangzhou Baiyun International Airport. The system can realize the design, production, installation, commissioning, acceptance, use and maintenance of the MEP equipment (systems) via an integrated management platform, and ensure economical use of energy, water and materials during the operation and maintenance of the Project. Information technology-based tools are used in property management to inspect and debug public facilities and equipment on a regular basis, optimize their operation according to monitoring data, and enable information sharing of the MEP systems.

CHAPTER IV

合和建筑观设计实践

INTEGRATED AND HARMONIOUS ARCHITECTURE：DESIGN PRACTICE

交通建筑

TRANSPORTATION BUILDINGS

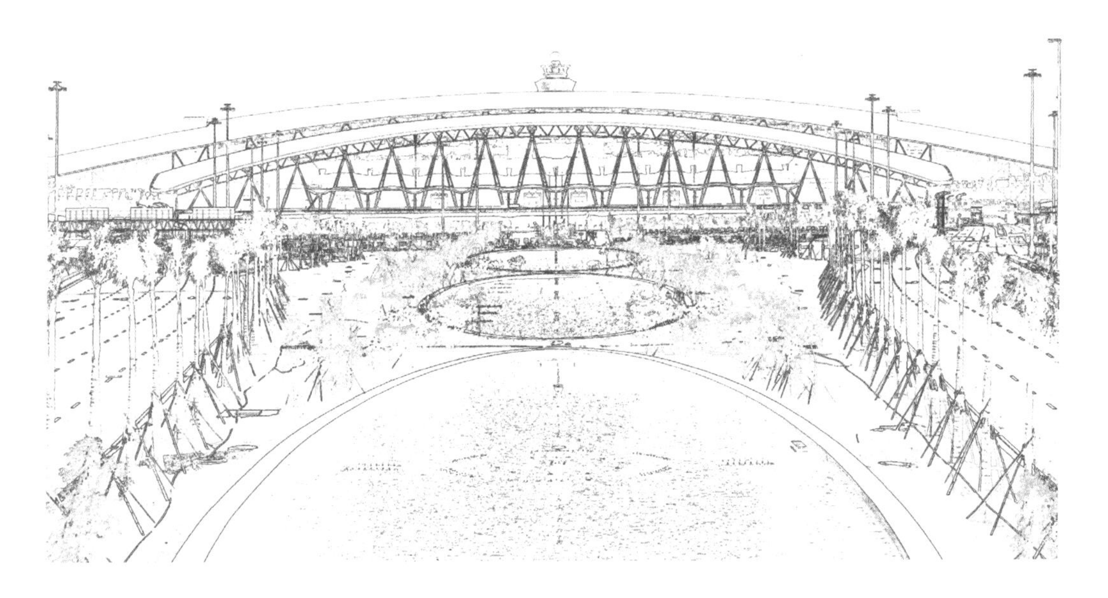

交通建筑是较为复杂的建筑类型，集多种交通方式、功能流线于一体，需进行要素平衡与多维融合，满足出行与体验需求，实现可持续发展，创造多元价值。

Transportation buildings are a relatively complex building typology that integrates different transportation modes and functional circulations. They need to balance various factors and realize multi-dimensional integration to achieve the desired passenger travel and experience, sustainable development and diverse values.

广州白云国际机场一号航站楼
Terminal 1, Guangzhou Baiyun International Airport, Guangzhou

项目地点：广州市，白云区
设计时间：1998- 2002 年
建设时间：2000- 2004 年
建筑面积：353,042 平方米
合作单位：美国 PARSONS 公司，URS Greiner 公司
建筑师团队（广东省建筑设计研究院有限公司）：
刘荫培、陈雄、郭胜、潘勇、周昶、李琦真、肖苑、吴华波、张宝棠、杨川、张展、潘伟江、李义波、沈钢、叶楠
主要奖项：全国优秀工程设计金奖
詹天佑土木工程大奖
"全国十大建设科技成就"称号
全国绿色建筑创新奖
中国"百年百项杰出土木工程"
中国建筑学会建筑创作大奖

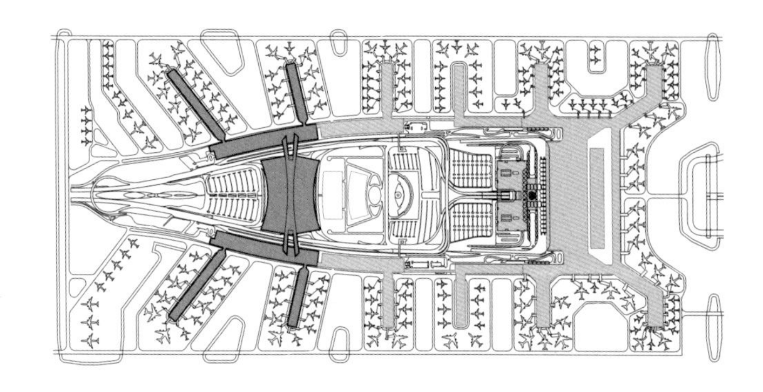

Location：Baiyun District, Guangzhou City
Design：1998-2002
Construction：2000-2004
GFA：353,042m²
Partners：PARSONS, URS Greiner
GDAD Design Team：Liu Yinpei, Chen Xiong, Guo Sheng, Pan Yong, Zhou Chang, Li Qizhen, Xiao Yuan, Wu Huabo, Zhang Baotang, Yang Chuan, Zhang zhan, Pan Weijiang, Li Yibo, Shen Gang, Ye Nan
Major Award(s)：The Gold Award of National Excellent Engineering Design Award, Tien-yow Jeme Civil Engineering Prize, The Top Ten Construction Technological Achievements in China, National Green Building Innovation Award, 100 Outstanding Civil Engineering Projects from 1900 to 2010 by China Civil Engineering Society, ASC Architectural Creation Award

白云国际机场迁建工程的一期项目从 1998 年起开始进行，持续到 2004 年结束，这 6 年以来的实践奠定了广东省建筑设计研究院在大型乃至超大型机场航站楼设计方面的基础。这个项目有一个非常重要的契机，就是在城市发展的过程中，广州的老机场需要搬到新机场，最终地点选在了广州白云和花都交界的北边，在这个契机下，业主方举行了一个国际竞赛。广东省建院当时想通过这个竞赛，积累大型航站楼建筑设计的经验，毕竟这种项目当时在国内是很少有的，这是我们从零开始的项目。广东省建院做机场的历史最早来自于 20 世纪 70 年代的汕头机场，到了 80 年代设计了老白云机场的国内候机楼，但这样大型、复杂的建筑项目，在广东省建院历史上也是从来没有做过的，1998 年是一个从零开始的过程。

当时业主方共邀请了 4 家航站楼设计的顶尖国外事务所，包括英国的 Foster、法国的 ADPI、美国的 Parsons 和 Greiner 联合体，以及加拿大 B+H 事务所。这几家在全球都是很顶尖的机场设计事务所，比如说 Foster 设计了香港机场，这是当时顶尖的国际机场；法国的 ADPI 在全球来讲是做机场最多的公司之一；美国的 Parsons 和 Greiner 在美国也做了许多机场。另外还有国内的三家设计院，当时都是各自独立参加，所以说广东省建院作为一方参与者，我们知道其实是没有办法中标的。即使这样，当时广东省建院的院长还是以非常独到的眼光，组织当时的领导班子开院务会讨论这个问题后，最后统一意见决定参加。因为当初如果不参与这个竞赛，我们后面将没有任何机会作为国内设计单位，我们希望通过参加这个比赛，跟国外优秀的同行学习怎样做这种大型复杂机场航站楼，同时通过这个过程，展示广东省建院的创新精神和服务态度。

当时我带领 10 个人的设计团队进行了三个月的投标，经历了从方案构思，到表达，到模型制作的一个完整流程。等到各方都提交成果后，业主方邀请了全国在机场设计领域比较有名的专家前来评审，出乎我们意料的是，专家们都觉得广东省建院的方案在国内几家投标单位中具有一定特色，整体表现最好。最终确定以美国的联合体 Parsons 和 Greiner 公司的方案作为中标方案，负责从方案设计到初步设计，同时确定了我们作为中方的配合单位，接续他们的初步设计，负责施工图设计和施工配合。

广东省建院除了在初步设计阶段与美国公司进行比较深入地配合，也在 1999 年去美国考察了几乎所有大型的机场，是一个很好的学习过程。在初步设计后期，1999 年的下半年还去美国公司进行配合。2000 年初初步设计评审通过之后进入施工图阶段，在此阶段广东省建院结合外方的方案和初步设计的成果进一步进行了优化，主要是各驻场单位和使用单位的一些需求落实得更加明确，包括海关、检验检疫、边检、航空公司、贵宾室流程、外事办等配合。因为外事办涉及省市政府需要进行的一些接待需求，这些需求也体现在设计里，所以在施工图阶段进一步地把这些主要需求进行融合，同时完成全部室内装修设计。项目到最后，2002 年到 2004 年的时候，在出完主要的设计图之后，主要的骨干到现场去驻场，驻场服务超过两年，在这个过程中基本上每天到工地开例会、工地巡场，及时地解决现场的技术问题，也克服了很多困难。这个项目当时在国内是非常有特色的，一是它的地铁直接就在航站楼的下方穿过，形成空轨无缝对接，这在国内是第一个，甚至直到现在还是唯一一个直接在航站楼正下方出入的地铁站，这种方便性是第一个特色。另外一个特色是从中轴线进去，也就是航站楼的南边，整个主立面就在机场路得到充分的展现。主楼对着机场路这样的一个布局，后来影响了中国很多其他机场的规划布局，包括北京机场 T3 航站楼、西安机场、深圳机场 T3 航站楼等，都是从机场路轴线进去，然后整个正立面就展现出来了。而之前很多机场都是从侧面看到航站楼，因为之前很多构型的主楼是平行机场路的，所以这也是一个非常重要的特点。整个设计还有很多先进的技术，比如幕墙技术、大跨度钢结构技术、大空间的空调技术等，这些在国内都是走在前列的。

在整个过程里面，业主、设计、施工，还有监理这四方充分配合，到最后呈现出来的完成度也比较高，当时在全国有很大的影响力。这个项目奠定了广东省建院在大型复杂机场航站楼这个领域一个很重要而且很高的起点。

白云机场迁建工程从 1998 年开始做总体的规划，1999 年航站楼总体规划，确定了一个有南北两个航站楼，以及多条指廊、双陆侧南北进场的总体布局。这个布局在大容量机场里面是比较重要的，南北贯通使得大容量的机场整个交通量的衔接以及辐射的区域会更多，

是比较重要的总体规划原则。从航站楼的角度来看，有几点特征后来也一直延续到 T2、T3 航站楼，包括了拱形的主楼、张拉膜的雨篷、“人”字形的结构柱。张拉膜雨篷跟岭南气候是非常契合的，因为是多雨炎热地区，日晒很充足，有了张拉膜雨篷之后旅客在车道边的出行体验会更加轻松，这个当时在全国是首创，并且直到目前在全国也是很少这样做。很多机场只有一个主楼的屋顶飘出来，没有车道边雨篷，在下雨的时候，或者猛烈的阳光下，雨篷可以遮阳挡雨，这种体验会非常好。这些特色反映了岭南气候，是设计回应气候的一些标志。

航站楼的外围护结构也非常有特色，是国内第一个在主体结构上面，也就是外围护结构上面使用张拉膜。除了雨篷之外，天窗也采用张拉膜。另外，T1 航站楼也是在国内比较早的大面积使用直立锁边铝镁锰屋面系统，虽然不是第一个用，但是这么大面积使用还是这个项目的首创。幕墙方面的特点更加突出，大面积地使用点式幕墙，在这里面点式幕墙还有很多种类型，包括鱼腹式自平衡桁架作为幕墙的二次结构，还有圆管二次结构，在不同的空间使用不同的类型。主楼用鱼腹式自平衡点式幕墙支撑结构，连接楼、空侧部分采用弧形圆管作为二次结构，指廊是采用竖向桁架作为二次结构。整个项目点式幕墙的使用在当时国内是面积最大的一个。点式幕墙主要的好处是给人感觉比较通透，创造出一个很通透的室内外的视野和空间体验，这在当时对全国的一些大型公共建筑，包括机场和体育场馆，产生了比较大的影响。

白云机场迁建工程构筑了一个人性化的、气势磅礴的、高科技风格的现代城市门户航站楼形象，为面向国际化迈进的广州，匹配了自身发展的新型大型国际空港。

Phase I of Guangzhou Baiyun International Airport lasted from 1998 to 2004. It was exactly those six years of practice that laid a solid groundwork for GDAD to build up competence in designing large and super large airport terminals. Before that, GDAD had already designed Shantou Airport in the 1970s and took part in the domestic terminal design of the old Baiyun Airport in the 1980s. But neither of them was comparable with Guangzhou Baiyun International Airport, the size and complexity of which was rarely seen in China back then. In the hope of filling in the gap, when the client announced the international design competition for the new airport, GDAD immediately decided to try and get involved in that remarkable task.

The client invited five top international teams specializing in terminal design, including the Britain-based Foster, the France-based ADPI, the consortium of the US-based Parsons and Greiner, and the Canada-based B+H. Foster designed Hong Kong International Airport, a top-notch one of its kind in the world at that time. ADPI was then among the few firms that had designed the most airport projects on the planet. Parsons and Greiner also have their fair share of airport design works in the US. Apart from them, three Chinese design institutes also independently joined the competition. That left a quite slim chance for GDAD to win the contract as an independent participant. Nonetheless, the then president of GDAD was quite visionary and convened a meeting, at which the leaders eventually reached a consensus on registering for the competition and trying to win as a supporting local design institute (LDI). We were hoping to learn how to build such a large and complex terminal from outstanding foreign peers and to demonstrate our creativity and professionalism while competing with our rivals.

After registration, we set up a 10-person team who, under my leadership, worked three months for the competition, going through the whole process from conception to design representation and model-making. The client then brought together renowned airport design experts to evaluate the submitted entries. To our surprise, they all gave us credit for featured design and overall performance compared with other local bidders. The client eventually adopted the opinion of the expert panel, awarding the schematic design (SD) and design development (DD) contract to the winner, the Parsons + Greiner consortium, and appointing GDAD as the LDI to carry the DD work forward to construction drawing design (CD) and construction administration (CA).

GDAD made every effort to support the DD work of the international partner. We visited almost all large airports in the US in 1999 and rendered face-to-face design cooperation in the second half of the year. After the DD deliverables were approved and the CD stage was initiated at the beginning of 2000, GDAD took over, optimized the DD submissions, and further specified the demands of all residing partners and the users. Once the main design drawings were completed, we assigned our key designers to reside in the site from 2002 to 2004, during which attending regular meetings and touring the site for timely resolution of all technical problems were basically a daily routine to them. The project was highly distinctive in China at that time. For one thing, the airport pioneered in realizing seamless air-rail connection and is still the only domestic airport that offers direct entry into the metro line right under its terminal. For another, the entire facade of the project is completely displayed along the Airport Expressway, greeting visitors arriving from the central axis south of the terminal. Unlike that, most airports built before had their main building parallel to the Airport Expressway, so visitors can only see the side of the terminal at first sight. The layout of the main building facing the Airport Expressway influenced the planning of many new domestic airports, including Beijing Capital International Airport (T3), Xi'an Xianyang International Airport and Shenzhen Bao'an International Airport (T3). In addition, the project features such technologies as curtain wall, large-span steel structure and large space air conditioning that are fairly advanced at home.

With perfect coordination between the client, the architect, the construction contractor and the construction supervisor, the project became a great success and gained a high reputation nationwide. It marked a major milestone in and added an impressive first record to GDAD's practice of large, complex airport terminal design.

The master planning for the relocation of Baiyun Airport started in 1998. It defined in 1999 the master layout of one south terminal, one north terminal, multiple piers, and a double landside south/north entry mode. The layout means a lot to a large-capacity airport like the project, as the north-south connection can greatly facilitate traffic flow and expand service coverage throughout the airport. Some features of T1, such as arched main building, tensioned membrane canopy and herringbone structural columns, are continued in T2 and T3. Among them, the tensioned membrane canopy fits very well the rainy and sweltering climate of Lingnan, allowing passengers to move at ease along the curbside. These features demonstrating the architect's response to the local climate was the first and remains a rarity in China. Most airports don't have any curbside canopy except the cantilevered roof of the main building.

The envelope structure of T1 is also distinctive, as it marks the first domestic application of tensioned membrane to building envelop and the first extensive use of the Al-Mg-Mn vertical edged roof. The curtain wall system of T1 is even more eye-catching. T1 boasts the largest point-supported curtain wall at home back then. A variety of point-supported curtain walls are involved, including those with fish-bellied self-balanced truss or circular secondary structures, as the specific space may require. Point-supported curtain walls excel in transparency. With transparent interior/exterior view and amazing spatial experience, they greatly influenced the design of many large public buildings in China, including airports and sports venues.

The project creates a people-oriented, hi-tech style and spectacular modern terminal, making it an impressive city gateway that well fits the city's ambition to go global.

广州白云国际机场二号航站楼

Terminal 2, Guangzhou Baiyun International Airport, Guangzhou

项目地点：广州市，白云区
设计时间：2006- 2016 年
建设时间：2013- 2018 年
建筑面积：658,700 平方米
顾问机构：美国 MA 公司，美国 L&B 公司等
建筑师团队：陈雄、潘勇、周昶、郭胜、赖文辉、郭其轶、易田、钟伟华、邓章豪、杨坤、吴冠宇、温云养、罗菲、许尧强、戴志辉、倪俍、董轩、金少雄、黎智立、黎运武
主要奖项：SKYTRAX“全球五星航站楼”认证
SKYTRAX“全球最杰出进步机场”
CAPSE 全球最佳机场
WBIM 国际数字化大奖
全国绿色建筑创新奖一等奖
行业优秀勘察设计奖 · 优秀（公共）建筑设计一等奖
中国建筑学会建筑设计奖 · 公共建筑一等奖
中国威海国际建筑设计大奖赛银奖

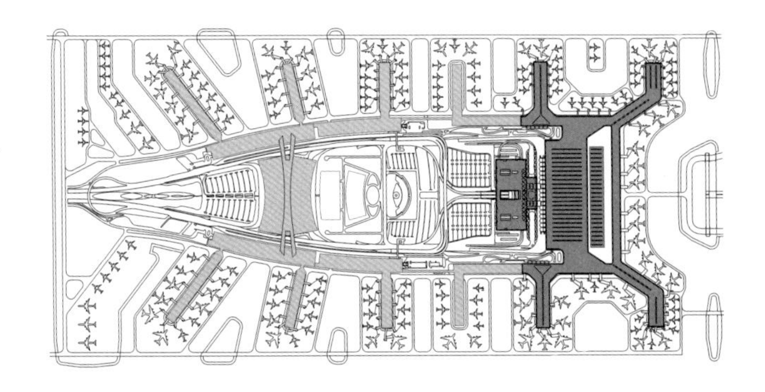

Location: Baiyun District, Guangzhou City
Design: 2006-2016
Construction: 2013-2018
GFA: 658,700m^2
Consultants: MA, L&B, etc.
GDAD Design Team: Chen Xiong, Pan Yong, Zhou Chang, Guo Sheng, Lai Wenhui, Guo Qiyi, Yi Tian, Zhong Weihua, Deng Zhanghao, Yang Kun, Wu Guanyu, Wen Yunyang, Luo Fei, Xu Yaoqiang, Dai Zhihui, Ni Liang, Dong Xuan, Jin Shaoxiong, Li Zhili, Li Yunwu
Major Award(s): SKYTRAX“Global Five-star Terminal”, SKYTRAX“The World's Most Improved Airport”, CAPSE The World's Best Airport, WBIM International Digitalization Award, The First Prize of National Green Building Innovation Award, The First Prize of Excellent (Public) Architecture Design of National Excellent Exploration and Design Industry Award, The First Prize (Public Building) of ASC Architectural Design Award, Sliver Award of the Weihai International Architectural Design Grand Prix

白云 T2 航站楼是在白云 T1 航站楼 2004 年投入使用以及 2009 年扩建之后，在2018年启用的一个航站楼，设计过程横跨了整整 12 年。该项目是一个对超大型复杂机场航站楼升级设计的探索，最早新机场是计划为 2010 年的亚运会服务，所以当时完成白云新机场转场之后，2006 年就开始设计 T2。后来经过业主的测算，认为扩建 T1 航站楼之后，再增加一些特殊航班的处理，运力是能够满足亚运会要求的。所以做了几年的方案到部分初步设计以后，该项目就暂停下来。2011 年重新启动了项目设计。这个历程有一个比较重要的转折点，在 2006 年到 2008 年那一段时间，整个 T2 还是遵循 1998 年时投标的国际竞赛方案，是南北两个楼分离站坪，而且是双陆侧，北边进入航站区也有陆侧，即 T1 是双陆侧，T2 也是双陆侧。

在 2011 年重启 T2 方案和初步设计的时候，综合了 T1 使用的情况，我们和空侧规划的顾问还有业主一起研究，提出了从分离站坪方案变到北站坪方案，将东、北、西三个方向的站坪连在一起，这样飞机的调度和站坪的使用效率会更高。该阶段还提出了国际国内的可转换概念，机位功能可以得到较大的提升。原来 T1 航站楼出发在主楼，到达在东西两边的连接楼，到达旅客的分离导致接客的人们会带来一定的不便。T2 新方案规避了该问题，采用主楼集中到达，与出发在同一个纵向空间内。总体来说，新方案站坪的效率更高，同时旅客的辨识度也更好。

机位布置上，在东、西、北三面环绕主楼的近机位，尽量把大机位的都布置在这些地方，旅客的平均步行距离就更短一些。流程布局上，有一个很大的改进和提升，其创造了当时全国最大面积的国内混流候机区，出发到达混流很大地提升了旅客的体验，旅客出发是由大厅到指廊，到达的旅客则经过指廊到大厅，避免了纯交通空间。混流指廊对于机场的资源也是一个很好的整合，包括商业、服务设施，同时面向出发和到达的旅客，这样一些设施的共享，使得资源更加高效。

基于白云机场地处中国南大门的区位优势，T2 航站楼拥有多条复杂的中转流程，除了始发终到这些点对点的航班之外，它还是一个国际航空枢纽，在楼内有国际转国际、国际转国内、国内转国际等多

种的中转航线。我们围绕航空公司，特别是以南航为首的天合联盟内的航空公司的需求进行设计，为他们枢纽航线的布局提供了非常适用的流程安排。这些中转的需求，从开始设计的时候，我们就与航空公司及机场公司进行了深入的对接，因此在后期的使用过程中，航空公司和机场公司都是非常满意的。这也说明一个问题，我们管理这种大型机场的设计项目，前期策划是非常重要的，尤其是与业主对接建筑的使用需求更加的重要。反映在机场项目的需求协调中，我们与联检单位的配合也更加成熟，在 T1 的基础上，联检的流程、空间的安排都得到了各方较高的评价。

从建筑设计的角度，T2 航站楼是非常有特色的。T1 航站楼的建筑形态没有特别强调明确的主题，到了 T2，则在设计的全程贯穿了“云”的主题，该主题的天空意向跟机场的航空功能是呼应的，同时和白云机场的区位命名也恰好契合。“云”的主题体现在张拉膜雨篷、主楼檐口的造型手法上，也体现在办票大厅的吊顶形式上，贯穿到办票岛、商业空间的设计，一直延伸到登机桥的形态。这其间都是动态的曲线和斜线，反映云的有机和流畅的特点。这些手法在 T2 航站楼内一气呵成地贯穿了各种空间和建筑细部，相对 T1 航站楼有了很大的提升。

T2 航站楼本身也延续了 T1 航站楼的一些特色，包括拱形的主楼形体和“人”字形柱的使用。T1 正立面使用的是跨度 18 米的“人”字形柱，因为 T2 尺度更大，放大到 36 米一个的“人”字形立柱单元。T2 的张拉膜雨篷也保留了与 T1 航站楼相似的基础元素。张拉膜雨篷不单呼应了云的造型，基于岭南气候的特点，张拉膜雨篷遮阳挡雨的功能也是机场的必备要素。主楼的吊顶系统也非常独特，这是一个原创的旋转渐变吊顶系统。通过吊顶叶片的渐变旋转和高低起伏，形成一个波浪型的室内界面，既可以让光透射进入航站楼，也遮挡到后面比较杂乱的网架结构，这也是我们在全国的首创设计。

在办票厅的中心设置了一个文化广场，可以作为文化展示和商业推广的用途，为办票厅营造了一个特别的公共空间并形成了焦点。办票岛系统加入了大量的自助托运柜台，在当时也是国内首个大规模使用自助托运行李系统的航站楼。在商业区跟旅客流程结合的角度，

我们以微型城市作为概念，在楼内收集了很多城市公共功能，包括商铺、餐饮、文化推送与展示，文化类产业同商业系统进行了有机结合，旅客流程同商业区深度融合，同时还将高端旅客候机布置在商业区中，并在商业区预留了未来进一步开发的空间。

内庭院跟航站楼的结合是一个重要特色，国内旅客过了安检，或者国际旅客过了联检后可以到达一组岭南风格的内庭花园，花园不单服务于高舱旅客的景观需求，也面向所有旅客开放。花园有植物绿化，有传统岭南建筑的坡屋顶组合，反映岭南园林的主题，这在国内机场里也是少有的地域特色表达手法。

登机桥系统围绕前述的航空公司枢纽需求，设置了 11 个可转换机位。这些国际国内可转换机位的流程设计，涵盖了所有的复杂类型，因此从可转换机位的技术探索角度，白云 T2 拥有当时国内乃至国际上最复杂、最先进的登机桥系统。

绿色环保设计方面，我们的目标最初就确定按国家绿色三星标准执行，从全专业的维度上围绕绿色三星进行设计，包括太阳能的使用、幕墙遮阳构件，也包括屋顶的深挑檐，还有雨水回收利用等新技术，这些都是基于满足绿色三星评价的要求。

在外围护结构方面，这一次我们主要使用了加强型铝镁锰屋面板系统，适应广州这种高台风地区的需求。幕墙方面，我们则相较 T1 航站楼作了一些调整，使用了横明竖隐的幕墙系统，垂直构件主要是玻璃间的隐藏吊杆，横向的型材框料则是本项目幕墙系统的特点所在，这些横向构件不单作为玻璃固定和幕墙第一道水平力传导路径，同时也是一个遮阳构件，同航站楼绿色三星的定位综合设计，尤其是航站楼西侧的阳光非常强烈，所以我们在西侧额外增加了可调节的电动遮阳百叶，是首个在全国航站楼内如此大面积地使用可调电动遮阳百叶的案例。

T2 航站楼内还预留了捷运系统的建设空间，包含捷运系统站点及线路高架，未来可以通过捷运系统与 T1 进行连接。

T1 航站楼是一个多方合作的设计作品，而 T2 航站楼则主要由广东省建院机场团队完成自主设计，在 2018 年投入使用后，得到了社会各界高度的认可，包括民航行业，也包括我们建筑行业内的专家和建筑师群体，以及社会大众方方面面都给予了高度的评价。这也是我们团队从 T1 开始一系列大型机场航站楼积累的经验升级，在 T2 厚积薄发所取得的成果。

T2 of Baiyun International Airport went into service in 2018 after T1 was put into use in 2004 and expanded in 2009. It took 12 years for the design team to explore how to upgrade the terminal building of a super large complicated airport. The design of T2 started early in 2006 not long after the relocation of the old airport, as the new airport was initially intended to support the 2010 Asian Games. Then the client found after calculation that the mere expansion of T1 plus the addition of some special flights already sufficed to meet the requirements of the Asian Games. Consequently, the work of T2 was suspended even after several years of schematic design and part of design development, until it was resumed in 2011. That process involved an important turning point. Between 2006 and 2008, T2 was designed as per the winning scheme submitted in 1998, which featured two separate aprons (one in the east and the other in the west) and a double landside transportation system.

After the design of T2 was resumed in 2011, we proposed the north apron scheme in place of the previous one with separate aprons after discussing with the airside planning consultant and the client based on the actual operation of T1. The new scheme links up the north, east and west aprons for more efficient aircraft scheduling and apron utilization. The idea of convertibility between international and domestic services was also proposed during that period, greatly improving the functionality of the stands. In T1, passengers depart from the main building but arrive at the east and west connecting buildings. The separation of arriving passengers makes it inconvenient for people to pick them up. The new scheme of T2 managed to avoid that problem

by enabling centralized arrival at the main building and keeping departing and arriving passengers within the same vertical space. In general, the new scheme is more efficient in apron utilization and more discernible to passengers.

As for the layout of stands, the main building is surrounded by large contact stands on the east, west and north sides, further reducing the average walking distance of passengers. The new scheme also significantly improves the process layout of T2 by creating the then largest waiting area in China mixing domestic departing and arriving flows, hence much better travel experience. The mixed flow piers also perfectly integrate airport resources for higher utilization efficiency. T2 also features a number of complex transfer processes in addition to point-to-point departure-arrival flights, because it also serves as an international aviation hub. We fully considered the demands of airlines, especially SkyTeam Alliance members led by China Southern Airlines, offering an easy-to-use process based on the layout of their hub routes. We had in-depth discussions with the airlines and the airport company over the transfer demands right at the beginning of our design, so the later service of T2 satisfied all of them. Through the design of T1, our cooperation with the joint inspection unit became smoother in T2, so the inspection process and space layout of T2 were highly praised by all. That also demonstrates the significance of initial planning, in particular the communication with the client about the use demands of the subject building, to the design management of airports in such a big size.

T2 is awfully distinctive in architectural design. The architectural form of T1 does not emphasize a clear theme. But in the case of T2, the theme of "cloud" runs through the whole design process. The sky imagery of the theme echoes the aviation function of the airport, and coincides with the location naming of Baiyun (meaning white cloud) Airport. The "cloud" theme is reflected everywhere from the modeling technique of the tensible membrane canopy and the cornice of the main building, to the ceiling form of the check-in hall, the design of check-in islands and commercial spaces, and the shape of boarding bridges. The dynamic curves and diagonal lines all resemble the organic and smoothly floating cloud. These approaches successfully connect through various spaces and components in T2, making it a tremendously upgraded version of T1.

Some features of T1 are also carried on in T2, such as arched main building, herringbone column, tensioned membrane canopy, etc. Among them, the tensioned membrane canopy responds to

not only the cloud theme in shape but also the climate of Lingnan in all-weather functionality. The suspended ceiling system of the main building is also unique. The originally created ceiling with gradually changing, rotating thus undulating blades forms a wavy indoor interface that can not only let light in but also hide the messy grid structure. It is the first of its kind in the country.

A cultural square is planned in the center of the check-in hall for cultural display and commercial promotion, creating a special public space and a strong visual focus. A large number of self-service check-in counters are added to the check-in island system, which was also the first in China. A good deal of urban public functions, such as shops, catering, cultural promotion and display, are provided inside T2 to create a micro city, where cultural industries are organically combined with the commercial system and the passenger handling process is deeply integrated with the retail area. The retail area also houses the VIP waiting area and the space reserved for future development.

The combination of the inner courtyard and the terminal is another important feature. After domestic passengers pass the security check or international passengers pass the joint inspection, they will be greeted by an array of Lingnan-style inner courtyard gardens that are open to all rather than just high-class passengers. The approach of using themed gardens to represent local features is rarely adopted by domestic airports.

The boarding bridge system has nine fixed ends for conversion between international and domestic services in response to the aforesaid airline demands. The internal process design of convertible stands covers all complex types. Therefore, from the perspective of technical exploration of convertible stands, T2 boasted the most complicated and advanced boarding bridge system at that time in China or even the world.

In terms of green building design and environmental protection, T2 was designed as per national three-star green building standard right at the beginning. That standard was implemented by all specialties and reflected by solar energy utilization, curtain wall shading components, deep roof overhang, etc..

The building envelope of T1 is highlighted by the reinforced Al-Mg-Mn roof panel system adapted to the climate of Guangzhou that is frequented by typhoons. With some adjustments to that of T1, the curtain wall system of T2 features a horizontally exposed and vertically hidden frame. Its vertical components are dominated by hidden suspenders between the glazing, while its horizontal profiles, as the key feature, can not only fix the glazing and conduct the first horizontal force of the curtain wall but also provide sunshade, echoing the three-star green building positioning. T2 marks the first extensive application of adjustable electric sun shading shutters to a domestic airport terminal. The shutters are installed in the west façade to block the burning sun.

Space for the APM system is also reserved in T2, including its station and elevated structure. The system is expected to be connected with that in T1 in the future.

While T1 is the product of multi-party design, T2 is essentially an independent design of GDAD's airport design team. Since its operation in 2018, T2 has been highly recognized by all sectors of society, including the civil aviation industry, experts and architects in the construction industry and the general public. T2 witnessed the upgrading of GDAD's expertise and experience in designing large airport terminals after T1.

广州白云国际机场三号航站楼

Terminal 3, Guangzhou Baiyun International Airport, Guangzhou

项目地点：广州市，白云区
设计时间：2019 年至今
建筑面积：450,000 平方米
合作单位：ADPI，SPS，金航诚规划设计有限公司
建筑师团队（广东省建筑设计研究院有限公司）：
T3 航站楼：陈雄、潘勇、赖文辉、李琦真、周昶、莫颖媚、潘玉婷、韦锡艳、何静、陈艺然、孙家明、詹明澄、方翔锋、黎嘉乐、梁景豪、姚粟尹、陈小东、夏伟豪、徐中天、邓章豪、伍思迪、翟卓越、王佳璇、张世挺、黎昌荣、廖志敏、高康、何业宏、金少雄、许尧强、温云养、陈业文、赵一飞、李桂芳、梁杏娟、周晓哲
GTC：陈雄、潘勇、陈奥彦、卢仲天、孙璐、王振鹏、陈卓宇、胡南江、肖晓苗、张俊威、黄智尚、赖伟健、陈毅恩、周靖葵、万晚霞、林振华、林琪琪、邱燕婷、石殷忆、崔洪亮、周瑜涛

Location: Baiyun District, Guangzhou City
Design: 2019 to present
GFA: 450,000m²
Partners: ADPI, Strategic Planning Services, Inc., Jinhangcheng Planning and Design Co., Ltd.
GDAD Design Team:
T3: Chen Xiong, Pan Yong, Lai Wenhui, Li Qizhen, Zhou Chang, Mo Yingmei, Pan Yuting, Wei Xiyan, He Jing, Chen Yiran, Sun Jiaming, Zhan Mingcheng, Fang Xiangfeng, Li Jiale, Liang Jinghao, Yao Suyin, Chen Xiaodong, Xia Weihao, Xu Zhongtian, Deng Zhanghao, Wu Sidi, Zhai Zhuoyue, Wang Jiaxuan, Zhang Shiting, Li Changrong, Liao Zhimin, Gao Kang, He Yehong, Jin Shaoxiong, Xu Yaoqiang, Wen Yunyang, Chen Yewen, Zhao Yifei, Li Guifang, Liang Xingjuan, Zhou Xiaozhe
GTC: Chen Xiong, Pan Yong, Chen Aoyan, Lu Zhongtian, Sun Lu, Wang Zhenpeng, Chen Zhuoyu, Hu Nanjiang, Xiao Xiaomiao, Zhang Junwei, Huang Zhishang, Lai Weijian, Chen Yi 'en, Zhou Jingkui, Wan Wanxia, Lin Zhenhua, Lin Qiqi, Qiu Yanting, Shi Yinyi, Cui Hongliang, Zhou Yutao

T3 航站楼是从 2019 年开始举行国际竞赛，在 2020 年，确定了我院联合体方案为中标方案。在 T1 和 T2 航站楼组成的白云机场第一航站区东侧，由 T3 航站楼及附属设施组成第二航站区。至此，白云机场形成了五条跑道、两个航站区、三个航站楼这样一个在全球机场中非常独特的总体规划。

T3 航站楼主要定位是为东航和国航定制的新航站楼，在白云 T2 建成后，中国民航这些年的发展是非常迅速的，因此到了 T3 航站楼必然会形成相对 T1 及 T2 航站楼的迭代及提升，社会各界对此的期待都会更高。在该项目上面，我们倾注了更多的创新，在设计团队方面，因为该项目设计为国际竞赛，方案的联合设计方包括航站楼部分的 ADPI、空侧部分为美国 SPS，以及国内的金航诚规划设计有限公司，联合体由我院牵头组织，最终获得中标。

T3 航站楼的方案有几个明显特点。首先是构型方面，它是一个在两条跑道之间的航站楼，航站楼两侧的跑道间距是 1600 米，而 T1 和 T2 所在的第一航站区两侧跑道间距是 2,200 米，所以 T3 不像第一航站区的 T1 和 T2 那样的多指廊布局，我们用了一种“H”形和“X”形综合变种的构型，更好地适应了航站区的跑道间距；该构型主楼周边的近机位比较多，四条指廊往南北两个方向进行延伸，旅客的平均步行距离较短；该构型所形成的 X 形岸线较长，空侧形成了包括东站坪、北站坪、西站坪的连续站坪，因此整个空侧效率较高。另一方面，该构型的南指廊可以预留下一期扩建空间，本期只建设主楼及北指廊部分，并保留在北侧建设卫星厅的可能性。设计联合体从各方面进行了充分地讨论和研究，最终得到了这样一个比较高效的航站楼构型。

第二是交通格局方面，如果说 T1 是国内第一个跟地铁密切联系的航站楼，那么 T2 增加了地铁及城轨的联系，到了 T3 航站楼，整个空铁联运的格局完全是一个更大的升级版。在该航站区内有两条高铁、两条城际外加地铁，同 T3 航站楼进行无缝衔接，穿越航站区的轨道同主楼是横向平行关系，在南侧楼前穿越交通中心和停车楼，交通

中心下面为高铁站、城轨站和地铁站。本方案的交通格局设计形成了空铁联运的最新一代产品，是一个重要的特色。

第三是概念主题方面，T3 航站楼相比原来的 T1 及 T2 航站楼有更鲜明的主题。T2 航站楼是以取自白云机场的“云”为主题，T3 航站楼的主题灵感则来自于花城、羊城，最终形成了“花开羊城 · 羊城花冠”的主题。这个主题体现在 T3 航站楼第五立面的造型上，其本身就是一个抽象提炼而成的花的造型。从建筑造型的角度，航站楼大空间主体结构柱采用了花瓣柱的造型，陆侧车道边立面、空侧立面也选取了花的形态，再组合行李提取厅等空间的花瓣柱元素，共同演绎出了羊城花开的主题。从各角度来说，T3 航站楼的标志性和主题性都有了更加明显的提升。

第四是航站楼流程方面，我们从航站楼的定位出发，T3 是为两个大型航空公司共同打造的枢纽航站楼，因此我们提出了双枢纽概念，国际流程在中间，两边是两家航空公司的国内流程，这样的流程关系，中转联程会高效便捷。这是一个围绕服务航空公司打造的枢纽流程布局，形成了 T3 航站楼的服务特色。

第五是航站楼室内氛围方面，与目前流行的很多航站楼不同，T3 是一个更加追求自然体验的航站楼，它是围绕花的主题展开的，在室内氛围方面扩展植物与自然的主题，并围绕这些元素进行的一些构建。在 T3 航站楼我们提出了一个新的概念——“去机场化”，我们希望旅客一进到航站楼，就跟常规机场一些常有的体验不同，它给人感觉更像是一个城市公共的门户或者客厅，同自然有密切的结合。航站楼东西两边的景观庭院，相比 T2 航站楼又是一个升级版，T2 的主流程没有经过室外庭院，而 T3 的主要流程就围绕在庭院周边展开。另外 T3 的室内基调基本上是以仿木为主，加上白色的流畅线条，配合花的主题。综合以上环境的特点，T3 航站楼能够有更加创新的旅客出行体验。

要特别强调的是，T3 航站楼的出发、到达和中转大厅是我们非常精

心打造的一个空间，非常出彩、非常有特色，充满阳光、充满植物的自然气息。特别是其中的文化广场，相比 T2 的文化广场得到了进一步的提升强化，T3 航站楼的文化广场面积更大，围绕文化广场的空间及景观打造也进一步提升，形成了 T3 航站楼的特色空间。

最后需要重点解析一下 T3 航站楼配套的交通中心，T3 航站楼同交通中心的连接关系继承于 T2 航站楼，陆侧的到港和出港流程都汇聚在交通中心一个楼内，但 T3 交通中心跟轨道的衔接有了进一步的提升。相比 T2 交通中心的第二个提升，体现在 T3 交通中心打造为羊城的新客厅，交通中心将商业娱乐、酒店、交通这些功能密切地整合在一起。

在交通中心的中庭里面我们提出了故乡水的概念，故乡水最早出现在 20 世纪 80 年代的内地，最早的五星级酒店——白天鹅宾馆，当时在全国乃至国际上都产生了很大的影响。可以想象这么一个情景，海外的游子回到祖国，回到羊城广州，看到故乡水概念景观，能感受到一份家乡对其欢迎的回应。我们希望“故乡水”概念主题，能够放在羊城的新客厅里，迎接海外归来的亲友，给市民带来温暖的体验。

GDAD consortium was awarded the design contract of T3 in 2020 through an international competition held in 2019. With T1 and T2 constituting the No. 1 terminal area and T3 and its ancillary facilities forming the No. 2 terminal area, a globally distinctive master plan with five runways, two terminal areas and three terminal buildings had taken shape in Baiyun Airport.

T3 is essentially customized for China Eastern Airlines and Air China. The rapid development of the aviation industry in China after the completion of T2 naturally made it a matter of course that T3 should be an upgraded version of T1 and T2, still less the higher expectations of all sectors of society. With ADPI working on the terminal design, the US-based SPS and the China-based JHC providing the airside design, the GDAD-led consortium eventually won the contract.

The design of T3 is distinctive in several ways. 1. Configuration. The configuration of T3 is a comprehensive variant of the H- and X-shaped ones. As the distance between the runways on both sides of T3 is 1,600 m while that of T1 and T2 is 2,200 m, the novel configuration of T3 can better adapt to the runway spacing compared with the multi-pier layout of T1 and T2. It allows for many contact stands around the main building and a shorter walking distance on average with the four piers extending to both the north and south. The resultant long X-shaped sideline forms continuous aprons in the east, north and west on the airside hence higher service efficiency. In addition, the configuration can reserve space for subsequent expansion in the south pier. The project only involved the main building and the north pier, with space reserved for building a satellite hall in the north.

2. Traffic pattern. With air-rail intermodality, T3 represents a big upgrade of T1 that took the lead in China to establish close terminal-metro connection and T2 that added intercity rail connection to the picture. Two high-speed rail lines, two intercity rail lines and one metro line are seamlessly connected with T3 in the terminal area. The rails traversing the terminal area are horizontally parallel to the main building, passing through the GTC and the parking building south of the main building. Under the GTC are high-speed rail, intercity rail and metro stations. The traffic pattern is a highlight of T3 representing the latest generation of air-rail intermodality products.

3. Concept theme. The theme of T3 is more distinct than that of T1 and T2. While that of T2 is "cloud", the design theme of T3 is “wreath of blooming Guangzhou" inspired by the nicknames “City of Five Rams" and “City of Flowers" of Guangzhou. Such theme is reflected in the fifth facade of T3, which takes an abstract flower shape. In terms of building components, the main structure of the large space of T3 adopts petal-shaped columns, while both the landside and airside facades simulate the shape of flowers. The petal-shaped column element is also seen in the baggage claim hall. Together, they echo the theme of “blooming Guangzhou". The landmark attributes and architectural theme are in all senses more evidently reflected in T3.

4. Passenger handling process. As T3 was intended as a hub terminal serving two large airlines, we put forward the concept of double hubs. The international process is placed in the middle, while the domestic process of the two airlines is arranged on both sides, enabling efficient and convenient connecting flight services.

5. Interior ambience. Different from many trendy terminals, T3 pursues a more natural travel experience, hence the interior theme of plants and nature. Therefore, we put forward a novel concept of "de-airport", hoping to make T3 feel more like a public city gateway or parlor. The eastern and western landscape courtyards make T3 an upgraded version of T2, because the main process of T3 is developed around them, while that of T2 involves no outdoor courtyard. In addition, the interior ambience of T3 is dominated by wood-like textures and some flowing white lines in conjunction with the flower theme. These characteristics combine to provide passengers in T3 a more innovative travel experience.

The departure, arrival and transfer halls of T3 are all meticulously designed spaces flooded with sunshine and natural plants. Its cultural square is larger in size and better in spatial and landscape creation than that of T2.

Last but not least is the GTC. The connection between T3 and the GTC is inherited from T2, i.e., both the arrival and the departure process at the landside are centralized in the GTC, except that the GTC is better connected with rail transit. Also, the GTC of T3 is designed as a creative city parlor integrating business entertainment, hotel and transportation. In the atrium of the GTC, we introduce the concept of hometown water, which first appeared at the White Swan Hotel, a five-star hotel in Guangzhou, in the 1980s, generating a great impact both nationally and internationally. Imagine the scene of an overseas resident returning to his hometown Guangzhou and being greeted instantly in the airport by the hometown water landscape in the new city parlor. Such a heartwarming experience is exactly what we look for through the design.

深圳宝安国际机场卫星厅

Satellite Hall, Shenzhen Bao'an International Airport, Shenzhen

项目地点：深圳市，宝安区
设计时间：2016- 2019 年
建设时间：2018- 2021 年
建筑面积：238,885 平方米
合作单位：Landrum & Brown，Aedas
建筑师团队（广东省建筑设计研究院有限公司）：
陈雄、周昶、罗志伟、陈艺然、林建康、李琦真、黄亦彬、李滔、伍思迪、吴杰明、赵一飞、王佳璇、李东、凌浩智
主要奖项：SKYTRAX“全球五星航站楼”认证
CAPSE 全球最佳机场
国际奖项协会缪斯设计奖铂金奖

Location: Bao'an District, Shenzhen City
Design: 2016-2019
Construction: 2018-2021
GFA: 238,885m²
Partners: Landrum & Brown, Aedas
GDAD Design Team: Chen Xiong, Zhou Chang, Luo Zhiwei, Chen Yiran, Lin Jiankang, Li Qizhen, Huang Yibin, Li Tao, Wu Sidi, Wu Jieming, Zhao Yifei, Wang Jiaxuan, Li Dong, Ling Haozhi
Major Award(s): SKYTRAX "Global Five-star Terminal", CAPSE The World's Best Airport , IAA Muse Design Awards (Platinum)

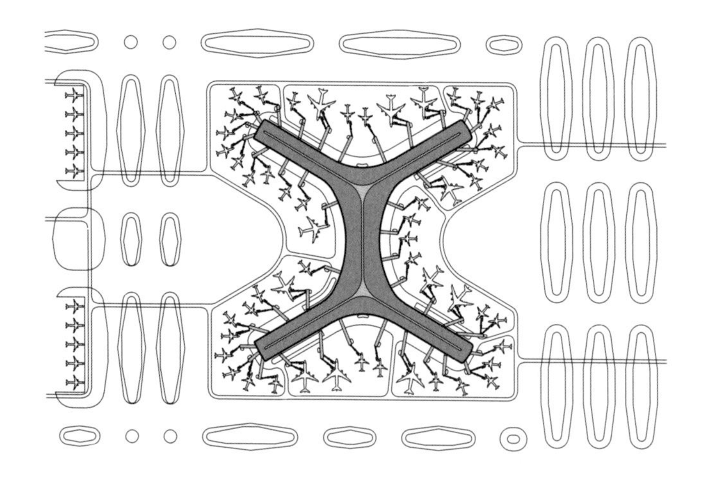

说到深圳机场卫星厅，不得不提到早年我们院跟 Fuksas 配合完成了深圳 T3 交通中心，以此为契机同深圳机场结缘。深圳机场卫星厅是一个国际竞赛招标项目，我院同 Landrum & Brown 和 Aedas 组成的联合体，获得中标并完成最终设计。从该项目 2016 年投标到 2021 年 11 月份投入使用，前后经历了超过 5 年的时间。

这个项目本身是比较有特色的，首先是构型方面。最早总体规划是“X”形，中间一个点向四面放射指廊，投标时我们觉得这个构型对站坪运行效率并不是最合适的，所以调整为了一个类似“X”形的变种，将原来“X”形中间的一个交叉点拉开了东西两个点，再分散出去形成四条指廊，构型中间连线作为捷运系统站点位置。这样的构型使站坪运行更加高效，与捷运系统衔接更加有序。同时还在两个节点打造了东西两个双子星商业文化广场。

项目从内到外都进行了全方位的探索，因为卫星厅主要是通过捷运系统为 T3 服务，因此从外形来讲，希望它跟 T3 航站楼能够对话，是一个和而不同的设计。T3 是以魔鬼鱼蝠鲼为概念做的海洋仿生学设计，以此呼应深圳作为海洋城市的身份，航站楼造型是蛋卷形，表皮非常有特色，是非线性渐变多孔状单元表皮；卫星厅作为后来者，我们希望它的外形和 T3 可以和而不同，首先我们的外壳也是蛋卷形的，也采用单元式建构了卫星厅的表皮；最终造型也形成了非常流畅的自然曲线过渡，包括登机桥都融入这种以流畅来表达海洋生物特性的主题中，以此同 T3 达成建筑语言的统一。但我们的表皮单元逻辑又与 T3 有一定差别，T3 是非线性渐变的，而卫星厅是规律性的表皮单元逻辑，其中非常有特色的“L”形遮阳构件，将绿色三星设计的需求融入表皮中，形成自己的特色。

因为塔台对卫星厅北边联络道有通视的要求，因此卫星厅北面两条指廊的高度是要压低的。这样导致蛋卷造型的高度也需要逐步地压低，这样蛋卷形的截面在不同位置都是非标的曲线，因此需要进行优化设计。

我们是怎么做的呢？我们把屋面优化成一个圆柱曲面，把它由指廊根部高点向指廊端部低点倾斜放置，两边的幕墙也是两个圆柱曲面，这样典型断面就形成了三段圆柱曲面的相交，两侧平行的圆柱曲面

被屋面曲率的圆柱曲面斜切相交，形成了外围护系统的三个主面，这样我们就利用圆柱固定曲率的特性，避免了使用三维曲面的问题。

卫星厅室内装修主题来源于深圳的城市性格，一是城市富有创新精神，产业科技高度发达，二是深圳有非常好的自然环境禀赋，因此室内设计强调科技和自然两个主题。另外我们提炼深圳市花三角梅和深圳机场 LOGO 的三角形元素，成为本次贯穿室内设计的母题。三角形母题贯穿了建筑—结构—装修，围绕母题进行一体化设计。从三角形结构单元构件的布置，开始呼应三角形的建筑母题；再到天窗下面的滤光器，我们用了三菱锥的方式去做，像钻石那种感觉，整个滤光器的布置跟旅客的流程是充分结合的；还有幕墙标准单元的外遮阳构件也是“L”形的类直角三角元素，贯穿了整个表皮设计；再就是三角形这种元素，天然就融入一些科技元素体中，比如说一些分子构成、晶体结构的线条，因此对三角元素的强调，达成了对科技主题的呼应。

卫星厅室内的自然主题，其一是通过仿木的吊顶来表达，采用三种深浅不同的转印仿木纹铝板进行吊顶的组合，天花吊顶的木纹色差使其更接近自然的质感，在消防规范对材料的限制下，尽量反馈了天然实木的感觉。

自然主题还反映在我们增加了立面的通透感，相对 T3 航站楼来说是一个重要的提升，T3 受实体表皮包裹了大部分外立面，很难做到对外视线的通透；卫星厅则在这个方向作了很大优化，整个外立面调整为比较通透的幕墙构件，视线的通透配合卫星厅内较多的室内绿化，使旅客的感觉更舒适和自然。卫星厅还有东西两个非常有特色的观景平台，位于东西主轴的两端，一边可看海，一边可观山，是室外小型的花园。旅客可以在室外的环境里面观山、看海、看飞机，获得非常特别的体验。

前文提到卫星厅构型所形成的东西两个商业文化广场也是本项目的特点，这两个广场都是三角形，三角形的三边都是由商业围绕在一起，而旅客的流程就经过这两个广场，因此两个广场的整体氛围非常活跃。两个广场中间的天窗也是三角形，我们在天窗下面吊装了环形荧幕，可以动态播出文化和商业信息。

卫星厅四条指廊以春夏秋冬作为主题进行设计，是非常有特色的，在全国机场也是首创。春、夏、秋、冬主要通过地毯、座椅、休息区，以及儿童活动设施的色彩来体现，系统化的色彩设计不单提高了指廊的辨识度及丰富度，也进一步提升了旅客的体验。

卫星厅中间的捷运系统站点也做了精心的设计，包括这些空间的打造，以及同商业的结合，在这些空间中我们还特别布置了一些裸眼3D屏幕，广告形态都不拘束于规则形体，体现出捷运站点空间追求突出、追求特色的特点。

卫星厅的地面装修材料选材也是该项目的一大特点，地面大量使用人造石铺地材料。这其中有两种人造石，一种是现场大面积现磨无缝人造石，另一种是现场铺装的预制人造石。卫星厅的核心空间捷运站点、中央的东西主轴、商业文化广场这些部分都是用大面积的现磨人造石，可以表现无缝的图案，我们在这里呼应了海洋和波浪的元素；指廊和到达走道的部分则使用了预制人造石。这是国内难得的大量使用人造石铺地的案例，一是为旅客创造了更丰富的体验，同时也体现出现代公共建筑环保节能的发展方向。

卫星厅将建筑艺术、实用功能、设计细节高度结合，打造一个具有高辨识度，属于深圳独一无二的卫星厅。

GDAD won the international design competition of the satellite hall of Shenzhen Bao'an International Airport in 2016 as a member of the consortium formed with Landrum & Brown and Aedas. It took five years of design and construction before the satellite hall was put into service in November 2021. Before that, we also designed the GTC of T3 of Bao'an International Airport in cooperation with Fuksas.

The project has many salient features. It adopts a variant of the originally planned X-shaped configuration, where the piers are radiated to all four directions from a central point. After we noticed that the original configuration was not the best choice for maximizing the operation efficiency of the apron, we split the central point into one east and one west point before the development of the four piers and placed the APM station at the central connecting line. The new configuration helped us win the competition for its capability of realizing higher operation efficiency of the apron and more orderly connection with the APM system. Its two nodes also house two commercial and cultural squares, one in the east and the other in the west.

Distinctiveness is reflected throughout the project from inside to outside. As the satellite hall was to serve T3 with its APM system, the design intended to establish a dialogue between them in appearance. T3 features the marine bionic design based on the concept of devil ray, echoing Shenzhen's identity as a marine city. The building takes the shape of an egg roll, with highly distinctive, non-linear, porous unit skin of gradual change. As a latecomer, the satellite hall is designed into the same egg roll shell and unitized skin. Its final shape also reveals a smooth natural curve transition, with the boarding bridge integrated into the theme of demonstrating the flowing characteristics of marine organisms. These measures combine to achieve the unity of architectural vocabulary with T3. Yet the skin unit logic of the satellite hall is somewhat different from that of T3. The former shows regularity while the latter non-linearity and gradual change. The L-shaped sunshade components incorporating the requirements of three-star green building design into the skin are also characteristic.

The requirement of the control tower for intervisibility into the linking taxiway to the north of the satellite hall means that the height of the two piers on the north should be lowered. As a result, the section of the egg roll indicates non-standard curves at different positions, which requires optimization. We optimize the roof into a cylindrical surface inclined from the high point at the root to the low point at the front end of the pier. The curtain walls on both sides of it also represent two cylindrical surfaces. The result is the intersection of three cylindrical surfaces on the typical section, which avoids using 3D curves leveraging the fixed curvature characteristic of cylinders.

The interior decoration theme of the satellite hall comes from the urban character of Shenzhen. The city is highly innovative, well developed in industrial technology, and blessed with abundant natural resources. Therefore, our interior design emphasizes two themes, i.e., technology and nature. Inspired by the triangle plum, the city flower of Shenzhen and the triangular LOGO of Shenzhen Airport, we employ triangular elements throughout architecture, structure and decoration, revolving our integrated design around the triangular motif. Examples include the arrangement of triangular structural unit components; the layout of the light filter in the shape of a triangular pyramid under the skylight that is fully combined with the passenger handling process; the L-shaped right angle-like triangle element of the external sun shading component of the standard curtain wall unit, which runs through the entire skin. As triangles are naturally integrated in some technological

EMPORIO ARMANI
TUMI

elements like molecular composition and crystal structures, the emphasis of the triangular elements also responds to the theme of technology.

The theme of nature inside the satellite hall is reflected by its wood-like suspended ceiling, which comprises three types of aluminum plates with transfer printed wood grain of different shades. The wood grain color difference feels super natural. Another reflection of the nature theme is facade transparency, a key improvement compared with T3. The facade of T3 is largely covered by the physical skin, making a transparent view into the outside hardly possible. To avoid that, relatively transparent curtain wall components are adopted for the facade of the satellite hall, where transparent view and abundant indoor greening guarantee a more pleasant and natural travel experience. The satellite hall has two small, highly impressive outdoor gardens, respectively at the east and west end. From these two viewing platforms, passengers can take a particular view of the mountains, the sea and the aircraft in an outdoor environment.

The aforesaid east and west triangular commercial and cultural squares are also an impressive trait of the project. Both triangles are surrounded by retail and covered by the passenger handling process that injects vitality into them. The skylight between the two squares is also triangular, with a circular screen suspended under it for displaying cultural and commercial information.
In the project, we also created the first four piers in China that are themed on spring, summer, autumn and winter by applying seasonal colors to differentiate their carpets, seats, lounge areas and children's play facilities. The systematic color design not only makes the piers more discernible and diversified but also improves the passengers' travel experience.

The APM system station in the middle is meticulously designed, including its space creation and combination with retail. Some naked-eye 3D screens are specially provided to free commercials from the regular form and highlight the pursuit of spatial prominence and characteristics in the APM station.

The extensive use of artificial stone for floor paving is also a major feature. The flooring is a combination of large area in-situ grinding seamless artificial stone and on-site paving of prefabricated artificial stone. The former is applied to the APM station, the central east-west main axis and the commercial and cultural squares, with seamless patterns responding to the elements of ocean and wave. The latter is applied to the piers and the arrival walkway. The massive use of artificial stone flooring is rarely seen in China, enriching the passenger experience while signaling the energy conservation tendency of modern public buildings.

The seamless combination of architectural art, practical functions and design details eventually makes a highly recognizable, unique satellite hall in Shenzhen.

深圳宝安国际机场新航站区交通中心
GTC, New Terminal Area, Shenzhen Bao'an International Airport, Shenzhen

项目地点：深圳市，宝安区
设计时间：2009-2010年
建设时间：2010-2012年
建筑面积：58,000平方米
合作单位：FUKSAS
建筑师团队（广东省建筑设计研究院有限公司）：
陈雄、罗志伟、肖苑、王梦、李建棠、邓弼敏
主要奖项：全国优秀工程勘察设计行业奖·建筑工程二等奖
广东省优秀工程勘察设计奖二等奖

Location: Bao'an District, Shenzhen City
Design: 2009-2010
Construction: 2010-2012
GFA: 58,000m^2
Partners: FUKSAS
GDAD Design Team: Chen Xiong, Luo Zhiwei, Xiao Yuan, Wang Meng, Li Jiantang, Deng Bimin
Major Award(s): The Second Prize (Construction Engineering) of National Excellent Engineering Exploration and Design Industry Award , The Second Prize of Excellent Engineering Exploration and Design Award of Guangdong Province

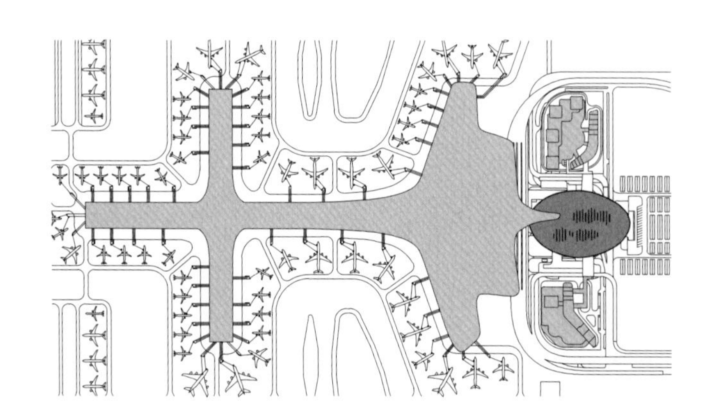

深圳机场GTC是为深圳机场T3航站楼配套建设的交通中心，2008年我院同Fuksas完成项目方案并获得中标，2013年交通中心落成投产。

深圳机场GTC作为深圳T3航站楼的配套设施，是国内最早的楼前独立地面交通中心案例。它主要是解决陆侧旅客尤其是到达旅客与其他地面交通工具的衔接问题，综合了陆侧的轨道交通、公交巴士以及小型社会车辆。

深圳机场GTC不单有陆侧交通衔接功能，同时引入了商业系统。进入GTC后，整个大厅是一个商业文化广场，大厅环境非常舒适，有自然采光，其装修元素由T3航站楼延续过来，采用了T3六边形的装饰概念母题作为天花的装饰单元。GTC大厅内也配置自然绿化，使得它有滨海城市的自然特征，提升旅客的出行感受。

深圳机场GTC的结构体系也非常独特，首先它是一个18米跨度柱网相对较大的公共空间，同航站楼的36米大跨是倍数关系，它的主体结构柱的铰接点在柱脚，因此根据弯矩传导特性，主体结构柱是上大下小。这种公共建筑内的非传统做法给人的体验很独特。

深圳机场GTC项目是广东省建院同深圳机场的第一个合作项目，除了深圳机场GTC以外，我院同期还负责了T3航站楼的前期各专业系统论证、设计监理及施工图审查工作，这也是深圳机场业主对我院白云机场项目经验的认可。

The GTC of Shenzhen Bao'an International Airport is the supporting transportation center of T3. GDAD won the design contract in collaboration with Fuksas in 2008, and the GTC was completed and put into use in 2013.

The project took the lead in building a free-standing GTC in front of an airport terminal in China. It mainly serves to connect landside passengers, especially arriving passengers, to ground transportation options including metro, buses, and small private vehicles.

In addition to enabling landside traffic connection, the GTC also houses a large commercial and cultural square with a pleasant daylit environment. The entire hall is decked out with the same hexagonal elements used in T3. Natural greening representing the coastal characteristics of Shenzhen is added to the mix to enhance the travel experience.

The GTC is marked by an extraordinary structural system. Its large public space is a product of an 18 m-span column grid system, the multiple of which is the 36 m-span one of T3. Its main structural columns are hinged at the foot, hence the larger upper part and smaller lower part given the bending moment transmission characteristics. Such an unconventional approach in a public building feels very different.

The project recorded GDAD's first cooperation with Shenzhen Bao'an International Airport. In addition to the GTC, we also provided the initial demonstration of various specialty systems, design supervision and construction drawing review of T3, which prove the client's high recognition of our work for Guangzhou Baiyun International Airport.

深圳机场东综合交通枢纽概念方案
Concept Design of Shenzhen Airport East Integrated Transport Hub, Shenzhen

项目地点：深圳市，宝安区
设计时间：2020- 2021 年
建筑面积：1000,000 平方米
合作单位：扎哈 · 哈迪德建筑师事务所，中国铁路设计集团有限公司，
深圳市市政设计研究院有限公司
建筑师团队（广东省建筑设计研究院有限公司）：
陈雄、郭其轶、罗若铭、梁智锋、倪俍、邓章豪、金少雄、钟仕斌、陈冠东、
龚锦鸿、郭文浩、陈铎文、卢宇

Location: Bao'an District, Shenzhen City
Design: 2020-2021
GFA: 1000,000m^2
Partners: Zaha Hadid Ltd., China Railway Design Group Co., Ltd, Shenzhen Municipal D&R Institute Co., Ltd.
GDAD Design Team: Chen Xiong, Guo Qiyi, Luo Ruoming, Liang Zhifeng, Ni Liang, Deng Zhanghao, Jin Shaoxiong, Zhong Shibin, Chen Guandong, Gong Jinhong, Guo Wenhao, Chen Duowen, Lu Yu

深圳机场东枢纽项目是我院同扎哈 · 哈迪德建筑师事务所、中国铁路设计集团有限公司、深圳市市政设计研究院有限公司作为联合体，于 2020 年年底参与的一项国际竞标，该项目也是我院再次尝试深度参与深圳机场的设计，我方联合体方案最终获得第二名。

深圳机场东枢纽的设计体现了深圳城市锐意创新的精神，该项目最大的亮点是将空铁联运的概念作为这一代空港交通综合体的产品理念，这是一次新的设计探索。集航空、高铁、城际、城市轨道、公交、出租车、小汽车等多种交通方式于一体，同时预留了枢纽同 T3 航站楼、卫星厅以及 T4 航站楼之间的捷运系统，因此枢纽的联运系统逻辑高度复杂，设计团队花了很多时间去思考及打造的一个空铁联运新模式。

我们的设计是如何达成这个目标的呢？首先枢纽地下有高铁、有城轨，一号线地铁则在高架上面穿过，地上部分主要是机场的 T1 航站楼。我们需要打造一个高效而强大的枢纽系统，不单完成航空之间的换乘，更需要达成航站楼和各种轨道交通之间的无缝衔接，同时要满足几种轨道交通之间的换乘，同时我们希望旅客的各种衔接换乘都有最好的体验。

最终我们研究的结果是通过一个扩大的空间进行无缝衔接联系。整个空间的基础是包括了高铁、城轨标准安检的管制区域，在经过了安检覆盖后，所有的轨道都可以通过这个扩大空间直接衔接，很方便地换乘。机场的管制区可以直接进入这个空间，这个扩大换乘空间进入机

场的管制区则需要增加一道升级安检，这样的话我们平衡了旅客在不同交通工具之间换乘需要反复进行不同等级安检以及安检管制区管理效率低的矛盾，实现了以一个标准安检扩大空间覆盖主要换乘流程的模式。在这个空间内旅客可以无缝衔接各种交通工具，而不需要反复经历安检，同时确保了空间内的管理便利和业态的丰富性。

空间营造方面，我们希望能把自然光线资源尽量引入地下的轨道交通候车空间，将轨道交通的候车大厅也打造成一个城市客厅，提升并强调轨道交通在整个交通枢纽中的核心地位。除了上述提到的扩大化共享换乘空间和高品质的地下轨道交通候车空间之外，我们在综合体的屋顶很有创意地设置了一个向市民开放的空间，往西南看，可以看到机场站坪的整个环境，包括大海、T3 卫星厅；往东北看，则可以通过一个巨大的屋面斜坡空间看到凤凰山，这是另外一个层面的城市客厅。

这个项目同时还是一个 TOD 开发项目，我们希望能将南北向的进场道路结合向东延伸到凤凰山的道路，形成十字轴线，将城市空间划分为南北地块，围绕枢纽进行一系列的开发，包含了产业的植入，城市公共空间的安排，以及城市景观的塑造，这些围绕枢纽打造的城市设计，也是枢纽综合设计的重要特色。

Shenzhen Airport East Integrated Transport Hub represents GDAD's another attempt to get deeply involved in the design of this airport in Shenzhen. The consortium formed by ZAHA, China Railway Design Corporation, Shenzhen Municipal Design & Research Institute Co., Ltd. and GDAD won the second place in the international bidding held at the end of 2020.

The design of the project mirrors Shenzhen's spirit of innovation. Its biggest highlight is the air-rail intermodality, a new exploration of airport transportation complex design. The project brings together various transportation options including aviation, high-speed rail, intercity rail, urban rail, bus, taxi and car, and reserves space for an APM system connecting the hub to T3, the satellite hall and T4. It took much time for the design team to work out the highly complicated logic of such a novel multi-modal system.

How to achieve this new pattern of transportation? The high-speed rail and intercity rail lines run under the hub, while Metro Line 1 passes over the hub. The above-grade part of the hub is dominated by T1. It takes more than connecting different airplanes to build an efficient and powerful hub system. Seamless connection between the terminal and various rail transit systems is a must, while the interchange between different rail transit systems is also necessary. We hope to secure a pleasant transfer experience for passengers.

The result of our research was to create an enlarged space for seamless connection. The whole space is a control area subject to the standard security check of the high-speed rail and intercity rail, and all rail systems can be directly interconnected via this enlarged space for easy interchange

between them. Direct entry from the airport control area to the enlarged space is made possible, but one needs to pass another upgraded security check process the other way round. This addresses the problems of repeated security checks at different levels for transfer between different means of transport and low management efficiency of the security control area. This enlarged space of standard security check covering all main transfer process saves passengers the trouble of undergoing repeated security checks, allows them to enjoy seamless transfer between various transport options, facilitates management and enriches business forms inside it.

In terms of space creation, the daylight is always maximized in the waiting space of underground rail transit, which turns the space it into an urban parlor and emphasizes the core position of rail transit in the entire hub. In addition to the above-mentioned enlarged transfer space and attractive waiting space of underground rail transit, we creatively provide a public space on the roof of the complex, where one can overlook the sea, the satellite hall and the entire apron to the southwest and Fenghuang Mountain to the northeast across a huge roof slope, hence a second urban parlor above the grade.

The project is also a TOD project. Our plan was to combine the north-south access road with the road extending eastward to Fenghuang Mountain, so as to form a cross axis that divides the urban space into a north and a south plot for an array of developments around the hub, including the implantation of industries, the arrangement of urban public spaces, and the shaping of urban landscape. All these urban designs around the hub also form important feature of our integrated hub design.

深圳机场 T4 片区规划及航站楼方案

Design Proposal of Terminal 4 and Terminal Area, Shenzhen Bao'an International Airport, Shenzhen

项目地点：深圳市，宝安区
设计时间：2020- 2021 年
建筑面积：424,010 平方米
合作单位：ADPi，SPS，金航诚规划设计有限公司
建筑师团队（广东省建筑设计研究院有限公司）：
陈雄、李琦真、陈超敏、陈艺然、翟卓越、陈伟根、王佳璇、李东、邓载鹏、邵国伟、许秋滢

Location: Bao'an District, Shenzhen City
Design: 2020-2021
GFA: 424,010m²
Partners: ADPi, SPS, Jinghangcheng Planning & Design Co., Ltd.
GDAD Design Team: : Chen Xiong, Li Qizhen, Chen Chaomin, Chen Yiran, Zhai Zhuoyue, Chen Weigen, Wang Jiaxuan, Li Dong, Deng Zaipeng, Shao Guowei, Xu Qiuying

深圳机场的 T4 航站楼（后更名为 T2 航站楼）项目是一个包括航站楼及周边航站区的设计，以国际竞赛的方式进行招标，我院联合了 ADPI、SPS 和金航诚组成设计团队，本团队是从白云 T3 航站楼延续下来的设计团队，因此配合非常默契，入围前三名。

项目有一个很重要的前置条件，在建筑场地下方有正在运行的轨道交通，同时本期也有轨道交通建设，因此如何在既有运行的轨道上以及正在扩建的轨道上面建设一个新的航站楼，是一个极大的挑战。经过业主组织的多轮论证，因用地所限，最终决定了航站楼需要架在多条轨道上面进行建设。之前很多航站楼是在轨道交通上面，但它们是同时建设的，该项目是国内首个在不间断运行的轨道线路上方施工建设的航站楼，无疑技术难度是更大的。除了工程管理经验之外，同时还要解决一系列的消防疏散问题，另外结构的托换也需要围绕直接上下叠置的轨道和航站楼之间的衔接来进行设计。航站楼还设置了捷运系统，将 T3、卫星厅、T4 串连起来。

在解决了上述的框架性问题之后，航站楼的构型设计也体现出我们的特色，我们同空侧顾问 SPS 和金航诚经过仔细分析之后，考虑到项目位于跑道的端部，其构型对整个跑道的运行效率会有直接影响，也就是说它的构型可以决定整个深圳机场的运行效率。经过充分论证后，我们选用了“Y”形构型，因为其他延展构型的航站楼在指廊端部的飞机推出时，飞机会直接影响跑道旁边的平行滑行道。当我们选用了收窄的“Y”形构型后，飞机推出之后仅需要经过航站楼两边的站坪，经过东西联络道才进入滑行道排队，大大减少了飞机推

出对平行滑行道内排队飞机的影响，这个根据空侧的特异条件确定的构型，是本项目的一大特色。

航站楼构型也给建筑本身确定了一个基本形态。深圳又名鹏城，深圳机场 T3 航站楼是海洋生物的形态，可称之为“鲲”，T4 航站楼的构型则是展翅的姿态，可称之为“鹏”，其一对应深圳的鹏城寓意，其二同 T3 组成南北鲲鹏的海天意向。

基于建筑构型的航站楼形态，有一个延伸向卫星厅的尾巴，还有一对向陆侧收窄耸起的翅膀，这个展翅的形象是非常突出的。为配合航站楼的整体形象，我们将羽毛作为构成屋顶造型的主要元素，形成了一个羽毛肌理的第五立面。立体的羽毛之间是航站楼屋面的侧向采光天窗，大面积铺设的屋面羽毛也使航站楼的室内天窗分布比较均匀，提高了自然采光质量，同时屋面的羽毛肌理通过天窗单元传导入室内，形成了富有主题的动态室内空间效果。鹏的主题也延伸到交通中心上面的景观中，一直以羽毛的肌理叠加延伸出来；同时，交通中心两侧的车道的形态，也有飞鸟的动态，再次呼应了本期项目“鹏”的标志性。

The international design competition of T4 (renamed T2 later) of Shenzhen Bao'an International Airport covered both the terminal and the terminal area. GDAD won the contract in collaboration with ADPI, SPS and JHC, the same consortium that won the design competition of T3 of Guangzhou Baiyun International Airport. For that reason, we cooperated quite well on the design of T4 and was ranked among the top three.

The project has an important given condition, i.e. the rail transit lines, including one under operation and the other to be built together with the project. This poses a great challenge for us, as the terminal is literally built on top of an operating rail transit line and the expansion of another line. Rounds of argumentation organized by the client led to the decision of building T4 above multiple tracks, given the awkward site area. There had been completed cases that put a terminal above a rail transit line, but the two were built synchronously. T4 is the first domestic terminal built above an operating rail line, which undoubtedly brought greater technical difficulties. In addition to project management, we also need to address a series of fire evacuation problems, and realize structure underpinning by revolving around the connection between the vertically stacked tracks and the terminal. An APM system is also planned inside T4 for connection with T3 and the satellite hall.

The configuration of T4 is also worth mentioning in addition to the above framework features. As the project is located at the end of the runway, its configuration directly affects the operation efficiency of the runway, or the entire airport. After thorough discussion with the airside consultant SPS and JHC and full demonstration, we chose a Y-shaped configuration specific to the airside

conditions of the project. In the case of other extended configurations, when an aircraft is pushed out of the pier, it can directly affect the use of the parallel taxiway beside the runway. With the narrowed configuration, the pushed out aircraft only needs to pass through the apron on both sides of the terminal and then enter the east-west linking taxiway before queuing at the taxiway, which greatly reduces the impact on the queuing aircrafts in the parallel taxiway. The Y-shaped configuration is a big highlight of the project.

The configuration of T4 also defines the basic form of the building. Shenzhen is also known as Pengcheng, namely the City of Roc. While T3 mimics the shape of Kun (an enormous legendary fish, which could change into a roc), T4 imitates the spreading wings of a roc, echoing the "City of Roc" and the Kun shape of T3.

Based on the Y-shaped configuration, T4 develops a tail extending to the satellite hall and a pair of eye-catching wings narrowing and rising towards the landside. To match the overall image of T4, feathers are taken as the main element of roof design, hence a fifth facade of feather texture. Amid the feathers are lateral skylights in the roof of T4. The extensive distribution of feathers in the roof allows for evenly distribution of indoor skylights hence better daylighting conditions. The feather fabric of the roof is continued in the interior via the skylight units, forming a dynamic indoor space effect with a distinct theme. The theme of "roc" also extends to the landscape above the GTC in the form of overlapped feather fabric. The shape of lanes on both sides of the GTC also resemble a flying bird, which again echoes the theme of "roc" of the landmark building.

珠三角枢纽机场（佛山新机场）方案

Design Proposal of The Pearl River Delta Hub Airport (Foshan New Airport)

项目地点：佛山市，高明区
设计时间：2021 年至今
建筑面积：370,000 平方米（近期）
合作单位：ADP Ingenierie，SPS，中铁上海设计院集团有限公司
建筑师团队（广东省建筑设计研究院有限公司）：
陈雄、罗志伟、李琦真、陈超敏、陈艺然、林建康、黄亦彬、邵国伟、陈伟根、翟卓越、王占勇、王佳漩、梁蕙茵、徐中天

Location: Gaoming District, Foshan City
Design: 2021 to present
GFA: 370,000m²(Recently)
Partners: ADP Ingenierie, SPS, China Railway Shanghai Design Institute Group Co., Ltd.
GDAD Design Team: Chen Xiong, Luo Zhiwei, Li Qizhen, Chen Chaomin, Chen Yiran, Lin Jiankang, Huang Yibin, Shao Guowei, Chen Weigen, Zhai Zhuoyue, Wang Zhanyong, Wang Jiaxuan, Liang Huiyin, Xu Zhongtian

佛山机场是我院牵头，连同 ADPI、SPS 以及中铁上海院组成的联合体进行设计，并获得中标的一个项目。本项目是一次规划、分期建设的空铁联运新机场。佛山机场的定位是非常高的，根据广东省规划“3+4+8”的机场布局，它同广州白云机场及深圳宝安机场并列为广东省三个枢纽机场，因此其在整个珠三角乃至大湾区都是非常重要的空铁联运枢纽节点。

在佛山机场的设计中，我们充分围绕佛山的历史文化、地域特色等要素进行创作，我们希望设计能够体现中国的文化自信。佛山是岭南文化的发源地，当地有非常多的岭南传统文化元素，我们在其中选取了几个主题，包括龙舟的元素，也包括粤剧元素以及南国醒狮等岭南元素。设计的目标是能够打造一个既有现代风格、现代功能，同时有很多岭南文化标志，能够全方位展现地域特色的航站楼。

佛山机场航站楼采用了一个比较独特的构型，机场的跑道间距是 2,200 米，预留了比较宽的航站区用地，在这样的用地条件下，一定是多指廊的构型，因此我们首创提出了双子星 6 指廊方案，其中 4 条本期建设，2 条作为远期预留。这 6 条指廊分别聚焦在东西两个中心广场空间，指廊以五角星发散的角度衔接在东西两个中心空间，形成了两个五星广场的特色空间，其中东边的是国际加国内共用的五星广场，西边则是纯国内的五星广场，这是佛山机场航站楼构型以及它形成的空间特色。

航站楼设计引入了传统木构建筑的梁柱概念，在天花及内外空间建构了一套木构体系，这些木构元素从车道边开始，一直沿到室内办票大厅和指廊空间，贯穿了整个旅客流程，因此航站楼本身是一个很现代的建筑，同时在其中也有一个传统木构的建筑体系，体现出传统与现代的结合。由木构体系勾勒出来的天窗系统，也强调了五星广场的中心空间，并沿五星角度向各个指廊发散，天窗顺着旅客流程布置，起到空间引导旅客的作用。

由两个五星广场连接的东西向中轴形成了一个非常富裕的中庭空间，在这里不单解决了旅客的商业、餐饮需求，还在东西两边布置了航站楼的安检功能。构型带来的中庭空间允许我们可以设置一个管制区外的、北向开放的城市客厅，这个空间可以面向陆侧市民开放，特别是可以看到站坪，也配套有商业和餐饮，这是国内最大的从陆侧区域可以看到飞机的观景广场，形成了佛山机场的重要特色。

本项目的陆侧也进行了整体设计，强调本地文化特色，其中，醒狮和龙舟的形态贯穿了陆侧的交通中心与酒店，下方的高铁站、城轨站，以及南侧的商业综合体，作为一个整体来表达地域文化概念。GTC也强调空铁联运的枢纽模式，其下方融合了高铁、地铁等轨道交通线路，并营造了一个空间能够提供方便换乘的服务，创造了多式联运的模式。

GDAD, as the leading party, partnered with ADPI, SPS and China Railway Shanghai Design Institute Group Co., Ltd. to win the bid for Foshan Airport, a new airport built under one-time planning and phased construction for air-rail intermodal transportation. As per the "3+4+8" airport layout planned by Guangdong Province, Foshan Airport, alongside Guangzhou Baiyun Airport and Shenzhen Bao'an Airport, is one of the three hub airports in Guangdong Province, serving as a pivot of air-rail intermodal transport in the Pearl River Delta and even the GBA.

The design of Foshan Airport centers around the history, culture, and local characteristics of Foshan to showcase China's cultural self-confidence. As the birthplace of Lingnan culture, Foshan is home to abundant elements of traditional Lingnan culture. Among them, the dragon boat, the Cantonese Opera, and the Southern Lion Dance are incorporated into the design to create a terminal with both modern style and function, as well as Lingnan culture and local characteristics.

The terminal has a unique configuration. As the 2,200 m runway spacing allows for a generous site for terminal development, a multi-pier configuration is definitely the most desired solution. So we creatively proposed the twin-star six-pier configuration, including four for current phase and two for long-term development. Concentrating in the two central square spaces in the east and west, the six piers connect two central spaces from each angle of five pointed stars to form two five-star squares. The east square is shared by international and domestic airlines, while the west one serves domestic airlines only, showcasing the spatial feature resulting from the terminal configuration.

With reference to the beam and column concept of traditional wood architecture, a wood

construction system is created at ceiling and interior and exterior spaces. These wood construction elements extend along the whole passenger circulation from the curbside to the check-in lobby and concourse spaces. The modern terminal incorporating a traditional wood construction system celebrates the combination of tradition and modernity. The skylight system outlined by the wood construction system also highlights the central space of the five-star square and radiates to each concourse along each angle of the star. The skylights along the passenger circulation help guide passengers.

The east-west central axis connected by two five-star squares delivers a spacious atrium to accommodate commercial and catering needs of passengers and host the security check function of the terminal on the east and west. The atrium makes room for a northward city parlor beyond the control area, open to landside citizens with commercial and catering facilities. With a view to airport apron, it is the largest viewing square in China where aircrafts are visible from the landside area, a prominent feature of Foshan Airport.

The landside is also designed as a whole to emphasize the local cultural characteristics. The forms of the lion dance and dragon boat are referenced in design throughout the landside transportation center, hotel, high-speed railway station and urban rail station below, and the commercial complex in the south, to fully showcase the regional culture. The transportation center emphasizes its function as a hub for air-rail intermodal transportation. It joins the high-speed rail, metro and other rail transit lines below grade, offers spaces for easy transfer and enables operation of multi-modal transportation.

珠海金湾国际机场 T2 航站楼
Terminal 2, Zhuhai Jinwan International Airport, Zhuhai

项目地点：珠海市，金湾区
设计时间：2018- 2019 年
建设时间：2019 年至今
建筑面积：191,300 平方米
合作单位：SPS
建筑师团队（广东省建筑设计研究院有限公司）：
陈雄、李琦真、周昶、陈艺然、宋永普、赖文辉、邓章豪、何业宏

Location: Jinwan District, Zhuhai City
Design: 2018-2019
Construction: 2019 to present
GFA: 191,300m²
Partners: SPS
GDAD Design Team: Chen Xiong, Li Qizhen, Zhou Chang, Chen Yiran, Song Yongpu, Lai Wenhui, Deng Zhanghao, He Yehong

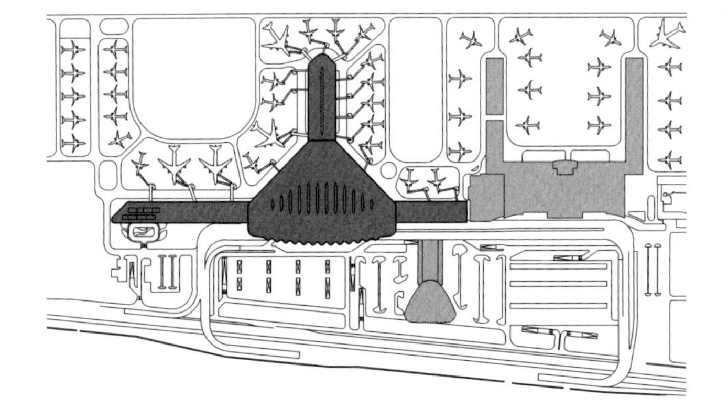

珠海机场是一个有很超前规划设计的机场，珠海 T1 航站楼的落成已经是在二十年前了，T2 航站楼则由广东省建院同 SPS 组的联合体在 2019 年参加国际竞赛并获得中标。

建筑部分是广东省建院的原创设计。珠海是一个很有地域特色的滨海城市，我们希望设计能够突出滨海城市的特征，体现海洋城市气质。因此该项目采用了贝壳和海浪作为设计元素，由此开展创作。珠海 T2 航站楼主体立面是一个柔美的贝壳造型，同前方的交通中心小贝壳相呼应，强调了滨海机场的特质。

珠海 T2 航站楼采用“T”字形的构型，在珠海机场当前的前置条件下，这样的构型有很高的站坪效率。首先珠海机场 T1 航站楼是双

指廊构型，若T2航站楼也采用双指廊构型，那么指廊间距就会很近，因此在当前发展容量的需求下，采用“T”字形构型有效地平衡了机位数量同站坪运行效率之间的关系。航站楼前设置了地面交通中心，交通中心内连接了城际轨道交通实现空铁换乘，轨道线路在楼前地下平行穿过。

基于“T”字形的构型，珠海T2航站楼的主楼和集中商业设施就天然集中在“T”字交点，在经过安检后，可以到达航站楼的商业空间，因为该商业空间非常紧凑，因此我们设置了多层的商业，很有城市商业空间的氛围，在机场商业内属于很特别的做法。考虑到珠海机场是中国航展的驻地机场，每年定期有航展及航空表演，其空侧景观资源在国内是非常独特的，我们在商业中心设置了空侧观景条件，以体现和利用珠海机场的特殊资源。

航站楼贝壳边缘海浪状的起伏，从室外檐口一直延伸进入主楼室内空间，形成室内空间的起伏单元，营造了室内空间的流动形态。在室内材质方面，我们将天花和地面都做成灰白纯色，以体现纯粹的贝壳及流动的形态。而在建筑内旅客平视范围，即室内立面，我们则采用了木纹质感的材料，以提高空间的亲和感与自然属性，并与纯粹的吊顶和地面形成对比，增加室内表皮的层次感。

珠海机场依托区域规划优势，打造“城市中的空港、空港中的城市”，为大湾区西翼再添航空新门户。

Zhuhai Jinwan International Airport is highly advanced in design, and its T1 was completed twenty years ago. The consortium of GDAD + SPS won the design contract of T2 through a competition in 2019.

The architectural design of T2 is the original work of GDAD. Zhuhai is a distinctive coastal city. To highlight its coastal characteristics, we employ shell and wave as the basic design elements. The main facade of T2 takes the shape of a mellow shell, echoing the GTC that appears like a smaller shell in front of it and emphasizing the characteristics of a coastal airport.

The T-shaped configuration of T2 ensures high apron efficiency with the given conditions of the airport. The repetition of the double-pier configuration adopted by T1 in T2 would result in very close distance between the piers. Therefore, to meet the capacity demand of the airport, we adopt the T-shaped configuration that can effectively balance the relationship between the number of stands and the operating efficiency of the apron. In front of T2, a GTC that allows rail-rail transfer is provided, with the intercity rail transit line running underground parallel to and in front of the terminal.

Naturally, the main building and commercial facilities of T2 are concentrated at the intersection of the T-shaped layout. The commercial spaces are placed right behind the security check area. Given the compact retail space, T2 is planned with multi-floor retail spaces which are commonly seen in cities but quite unusual for a terminal. The retail center is also designed as an airside viewing deck for people to watch the airplane exhibitions and air shows that are regularly held in the airport, fully reflecting and leveraging the unique resources of the airport.

The wavy shell edge of T2 extends from the outdoor cornice all the way into the main building, forming undulating units of interior space that flows smoothly and freely. As to interior materials, the ceiling and flooring are all in sheer gray and white to reflect the pure color and mellifluous shape of shells. The interior façade at eye level employs materials of wood grain texture to foster a more friendly and natural interior space and enrich the hierarchy of its interior skins through contrast with the purity of the ceilings and flooring.

Thanks to the advantages specified in the regional planning, Zhuhai Jinwan International Airport will create "an airport in the city and a city in airport" to add a new aviation gateway to the west wing of the GBA.

揭阳潮汕国际机场航站楼
Jieyang Chaoshan International Airport Terminal, Jieyang

项目地点：揭阳市，榕城区
设计时间：2007- 2010 年
建设时间：2008- 2011 年
建筑面积：58,752 平方米
合作单位：Landrum & Brown，上海新时代机场设计有限公司广州分公司
建筑师团队（广东省建筑设计研究院有限公司）：
潘勇、陈雄、肖苑、谭杨威、赖文辉、郭其轶、黄亿万、黄珏欣、陈颖、张春灵
主要奖项：全国优秀工程勘察设计行业奖 · 建筑工程公建三等奖
香港建筑师学会两岸四地建筑设计大奖卓越奖
广东省优秀工程勘察设计奖 · 工程设计二等奖

Location: Rongcheng District, Jieyang City
Design: 2007-2010
Construction: 2008-2011
GFA: 58,752m^2
Partners: Landrum & Brown, Shanghai Civil Aviation New Era Airport Design & Research Institute Co., Ltd.
GDAD Design Team: Pan Yong, Chen Xiong, Xiao Yuan, Tan Yangwei, Lai Wenhui, Guo Qiyi, Huang Yiwan, Huang Juexin, Chen Ying, Zhang Chunling
Major Award(s): The Third Prize (Construction Engineering - Public Building) of National Excellent Engineering Exploration and Design Industry Award, Nominated Award of Hong Kong Institute of Architects (HKIA) Cross-Strait Architectural Design Symposium and Awards, The Second Prize (Engineering Design) of Excellent Engineering Exploration and Design Award of Guangdong Province

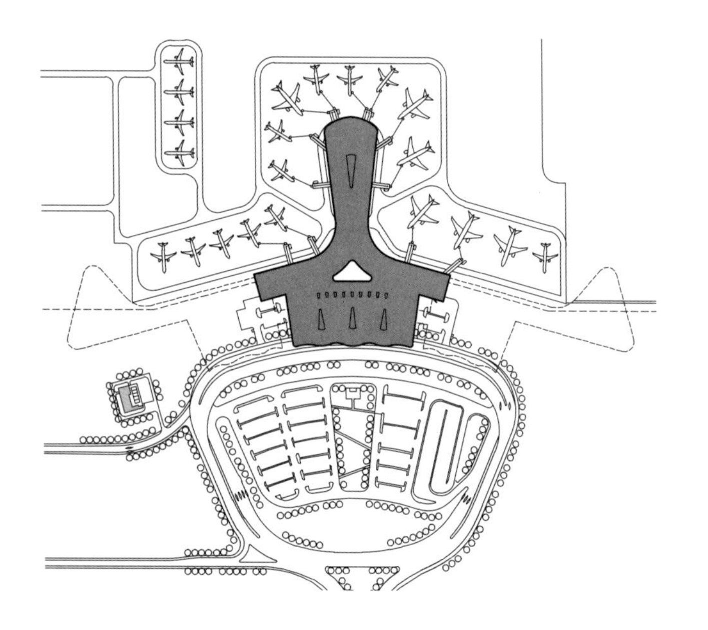

揭阳潮汕机场项目源自于 2007 年我院参加的一个国际竞赛，同兰德隆布朗和新时代广州公司合作投标，我院负责的建筑部分以原创方案中标。

航站楼的构型是非常有特色的。首先，航站楼设计的近期年吞吐量约 450 万人次，二期需扩展到 1,450 万人次，针对这样的分期需求，我们选用了中指廊的构型，这种构型最大的特色是既可以扩建指廊，也可以扩建主楼，从分期建设的角度来说非常灵活，并且扩建之后仍然是一个整体，对旅客来说还是一个独立的航站楼，非常方便。

航站楼的整个创作围绕着几个地域文化元素，一个是潮汕滨海的海洋元素；另一个是潮汕地区传统建筑群落的元素。航站楼屋面檐口是波浪形的构成，一方面是体现海洋元素，同时波浪的排列也体现了潮汕传统民居聚落排列方式，反映了当地传统建筑群落的韵味。

主楼跟指廊由一个一体化的大屋面覆盖，是一个非线性的屋面系统。远期扩建后，三条指廊的屋面形成一个“山”字构形，前端的三段主楼则是一个三点水的概念，三点水加上山字刚好组合出潮汕地区的“汕”字，这也是航站楼远期构型的寓意之一。

我们希望在揭阳潮汕机场这样规模的航站楼中，创造一个花园航站楼的模式，因此在航站楼的正中设置了一个中庭花园，花园位于办票大厅的正后方，旅客一迈入办票大厅，透过前列式的柜台，就可以欣赏到花园的景色。除了面向办票厅，旅客安检之后的流程直接围绕着这个花园展开。中庭花园一路贯穿到航站楼下层，因此无论是旅客到达走道，还是行李提取厅，都可以看到这个花园。贯穿整个航站楼的花园概念，让这样一个中小型航站楼有了自己很独到的特性，给旅客带来非常愉悦的感受。

In 2007, GDAD partnered with Landrum & Brown and New Era (Guangzhou) to participate in the international competition of Jieyang Chaoshan International Airport and won the contract with our original architectural design scheme.

The configuration of the terminal is very distinctive. The designed annual passenger throughput of the terminal is about 4.5 million in the near future and 14.5 million after Phase II is put into use. In view of the phased development needs, we adopt the central pier configuration. Its biggest feature is that it allows the expansion of both the pier and the main building thus is highly flexible from the perspective of phased construction. Furthermore, the terminal will remain an independent whole even after the expansion without causing any inconvenience to passengers.

The architectural creation of the terminal centers around regional cultural elements, such as the marine culture of coastal Chaoshan and the traditional housing settlements in the area. The wavy roof cornice of the terminal reflects both of them, including the layout and appeal of indigenous architectural clusters. The main building and the piers are covered by one large non-linear roof system. After the long-term expansion, the roofs of the three piers will feature an epsilon configuration resembling the Chinese character " 山 ", while the three main buildings at the front end are in a layout resembling the radical" 氵 ". Together, they will form the Chinese character " 汕 "representing Chaoshan. This is one of the implications of the long-term configuration of the terminal.

We hope to create a garden terminal in an airport the size of Jieyang Chaoshan International Airport. Therefore, an atrium garden is planned in the center of the terminal, right behind the check-in hall. Passengers can enjoy the view of the garden through the linear counters instantly after they step into the check-in hall. The passenger handling process after security check immediately revolves around the garden. The atrium garden extends all the way down to the lower floors of the terminal, hence is visible from both the passenger arrival walkway and the baggage claim hall. This endows unique identity to the terminal, impressing passengers in a rather pleasant way.

揭阳潮汕机场航站区扩建

Terminal Area Expansion, Jieyang Chaoshan International Airport, Jieyang

项目地点：揭阳市，榕城区
设计时间：2019- 2022 年
建设时间：2020 年至今
建筑面积：52,600 平方米
建筑师团队：陈雄、潘勇、郭其轶、邓章豪、温云养、许尧强、倪俍、金少雄、韦锡艳、江祖平、黄玉青、谢金龙、岑家裕、陈伟根、凌浩智

Location: Rongcheng District, Jieyang City
Design: 2019-2022
Construction: 2020 to present
GFA: 52,600m^2
Design Team: Chen Xiong, Pan Yong, Guo Qiyi, Deng Zhanghao, Wen Yunyang, Xu Yaoqiang, Ni Liang, Jin Shaoxiong, Wei Xiyan, Jiang Zuping, Huang Yuqing, Xie Jinlong, Ceng Jiayu, Chen Weigen, Ling Haozhi

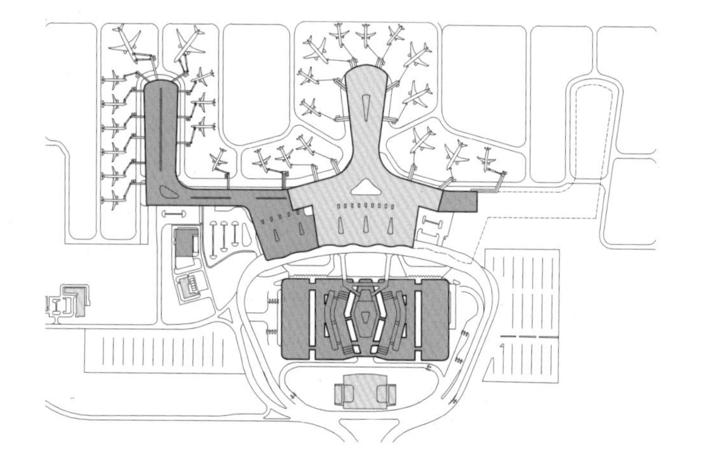

揭阳潮汕机场一期是从 2007 年开始设计到 2011 年投入运营，2019 年，机场开始研究扩建，我们的设计是在原来一期工程总体规划的基础上，扩建南侧国内主楼、国内候机指廊以及北侧局部国际候机指廊，扩建后的构型基本延续了一期规划设计的框架。

本期扩建工程主要是解决国内旅客流量的增加，扩建南侧主楼延续了一期航站楼波浪形屋面的设计语言，将来北边扩建主楼后，可以形成一个连续而对称的形态。指廊的部分也顺应了一期工程的一些做法，同时根据行业趋势做了一些调整，最大的调整就是把指廊的国内部分做了混流处理，调整后的混流指廊同原来的分流指廊也进行了很好的衔接。北指廊还是维持原有的国际功能，并适当延长进行了服务能力的提升。

本期航站楼延续了原来花园航站楼的主题，在新旧主楼的扩建衔接部分也增加了一个楼内中庭花园，一路延续布置在办票及安检的旅客流程上，确保了原航站楼旅客体验的亮点，呼应了原有的设计主题。除了航站楼本身的扩建，本期扩建另外一个最重要的内容是大大地提升陆侧的功能，扩建了交通中心综合体，将航站楼、交通中心、停车楼、已经建成使用的高铁站，通过中轴线的空间串联起来，同时为正在建设的城轨预留了衔接条件。交通中心的建设使这个中小型机场实现了空铁联运，且本项目的特别之处在于，对高铁的缝合衔接是利用现有投入运营的高铁站房，作为空铁联运衔接核心的地下通道，需要在不停航施工的条件下去完成，体现了航空类扩建工程的特色和难点。

扩建工程的交通中心是一个综合体建筑，地下通道和首层通道有集中商业空间，上层则布置了一个酒店，同时衔接了航站楼、高铁站以及城轨站点，对于旅客来说非常高效方便。体量稍高的旅客过夜用房也有意将航站区的中轴线空余出来，两边分开，形成一个两侧围合的构型，通透的中轴使到航站楼的空间形态得到一定的展示，强调了轴线感，这也是陆侧部分重要的设计逻辑。

Phase I of Jieyang Chaoshan Airport was designed in 2007 and put into service in 2011. Its expansion plan was put to discussion in 2019. Our design proposed to expand the main building and the domestic departure pier on the south and part of the international departure pier on the north on top of the Phase I master plan. The expanded configuration basically continues the planning framework of Phase I.

The expansion was mainly intended to accommodate the increased domestic passenger traffic. The southern expansion of the main building continues the wavy design vocabulary of T1, and will form a continuous, symmetrical whole with its northern expansion in the future. The pier expansion also continues the design approach of Phase I, with some adjustments made to keep up with the industry trend. Among them, the biggest adjustment is to adopt a mixed flow process in the domestic departure part of the pier, which is well connected with the original separate flow pier. The North Pier remains to serve international flights but is appropriately extended for higher service capacity.

The new terminal continues the original theme of garden terminal, adding an atrium garden to the transition between the new and the old main building. The garden extends all the way along the check-in and security check process, securing the pleasant passenger experience highlighted in the original terminal design and echoing the original design theme.

Another important aim of the project was to significantly improve landside functionality. The expanded GTC complex links up the terminal, the GTC, the parking building, and the high-speed rail station in service via the central axis, and reserves conditions for connection to the intercity rail under construction. The GTC enables air-rail intermodality in such a small/medium-sized airport. What's more, the project's connection to the high-speed rail is realized via the high-speed rail station building in use. That is, the underground passage essential to the air-rail intermodal transportation was built without suspending the transportation services, an impressive embodiment of the characteristics and difficulties of airport expansion.

The GTC of the expansion project is a complex with commercial spaces centralized in underground and first floor passages and passenger accommodation for overnight stay (hotels) placed on the upper floor. The latter is connected with the terminal, the high-speed rail station and the intercity rail station, which can greatly facilitate passenger mobility. These rooms in a bit higher block also intentionally free up space on the central axis and form a configuration enclosed on both sides. The transparent central axis, to some extent, enhances the presence of the spatial form of the terminal and emphasizes the sense of axis, an important design logic of the landside part.

湛江吴川国际机场航站楼
Wuchuan International Airport Terminal, Zhanjiang

项目地点：湛江市，吴川市
设计时间：2017- 2019 年
建设时间：2019- 2021 年
建筑面积：61,900 平方米
建筑师团队：陈雄、陈奥彦、宋永普、杨展辉、卢仲天、孙海音、张栋、龚哲、肖晓苗、黄智尚、张俊威、孙璐、林振华、万晚霞、王振鹏
主要奖项：广东省注册建筑师协会广东省建筑设计奖 · 建筑方案奖公建类二等奖

Location: Wuchuan City, Zhanjiang City
Design: 2017-2019
Construction: 2019-2021
GFA: 61,900m²
Design Team: Chen Xiong, Chen Aoyan, Song Yongpu, Yang Zhanhui, Lu Zhongtian, Sun Haiyin, Zhang Dong, Gong Zhe, Xiao Xiaomiao, Huang Zhishang, Zhang Junwei, Sun Lu, Lin Zhenhua, Wan Wanxia, Wang Zhenpeng
Major Award(s): The Second Prize (Architectural Design - Public Building) of the Guangdong Architectural Design Award of Guangdong Province Registered Architect Association

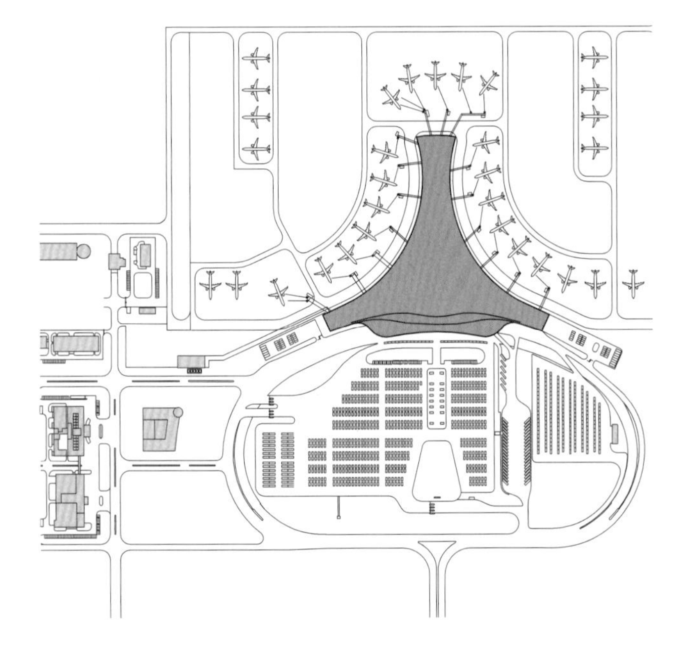

湛江机场是我院同兰德隆布朗在 2017 年参与设计投标，并获得中标的一个项目。该项目为迁建工程，由老的湛江机场迁建到吴川，项目于 2022 年转场并投产运营。

航站楼的规划构型为“人”字形，本期建设为中指廊的模式，这是基于湛江机场一期工程 510 万的年旅客流量运营规模所决定的构型。“人”字形构型是由主楼直接放射出三个指廊，旅客的平均步行距离比较短，同时机位的运行也比较方便，指廊布局不会出现“U”字形的港湾，两边都是大港湾，可以提供优质高效的空侧资源。

航站楼的整体造型基于湛江滨海城市的特征，采用较为流畅的建筑语言诠释，从各个角度看航站楼都是通过一些流畅的曲线勾勒出来的，其中主楼由三条曲线构成，一直延伸到指廊，通过曲线的组合，

给航站楼的整体造型引入了流畅而丰富的节奏感。值得注意的是，其中一条曲线将车道边雨棚结合在一起，不单解决了多雨地区高架桥出港体验的问题，还不同于一般机场分离式雨棚的做法，将车道雨棚做成和航站楼造型一体化，这是湛江机场航站楼正立面的一个重要特色。

有意思的是，湛江机场的竣工图片发布至网上后，网友根据航站楼的构型和平面曲线特色，给湛江机场亲切地起名为“最牛机场”，来形容航站楼类似一个“公牛头”的造型，这也是设计团队未曾设想的，反映了建筑使用者同建筑设计者之间的有趣互动。

新迁建的湛江机场位于湛江下辖的吴川市，在湛江和茂名之间，定位于服务粤西片区，作为粤西的新门户。粤西地区有着非常独特的地域文化，例如原始味道非常浓郁的红土文化，源自当地红土地的地质特性，其中广泛地包含了篝火舞、舞狮舞龙这类当地传统的文化元素，因此我们希望可以将当地的地域文化特色融入航站楼的空间中。

我们将红土文化的红色象征引入室内空间。例如航站楼室内座椅等彩色元素，还比如天花吊顶缝隙透露出的红色，反映了当地地域文化的背景底色。航站楼内大空间天花采用的是单元式的吊顶，每一个吊顶单元都是一个菱形，在每个单元增加了吸音需求的穿孔后，菱形一开二就变成两个三角形，又增加了一种表情元素，菱形加上三角形的元素在大天花上排列成了鳞片的海洋，呼应了地域的海洋特性。天花鳞片在两侧逐级跌落后，将鳞片的缝隙逐步放大张开，逐渐透出其间的红色，彰显了当地的地域文化底色。

GDAD partnered with L&B to participate in the bidding and win the design contract for Zhanjiang Airport in 2017. Relocated from its former location to Wuchuan, the airport was put into operation after a successful transition in 2022.

The herringbone-shaped terminal features middle piers to accommodate 5.1 million annual passenger traffic of Phase I. The main building directly radiates three piers for short average walking distance and convenient operation of aircraft stands. The resultant large bays instead of U-shape bays can ensure high-quality and efficient airside resources.

The overall shape of the terminal celebrates the characteristics of Zhanjiang, a coastal city, with smooth architectural language. From every standpoint, the terminal is outlined by sleek curves. The main building is composed of three curves extending to the piers. The combined curves bring fluid and rich rhythm to the overall shape. Notably, one of the curves joins the curbside awning, improving the departure experience of passengers arriving from viaduct in rainy areas and differentiating from separate awning commonly seen at airport. The curbside awning integrated with the terminal makes a difference to the terminal frontage of Zhanjiang Airport.

After the as-built photos of Zhanjiang Airport were released on the internet, netizens sweetly named the airport "the most bull-like (terrific) airport", as the terminal looks like a bull's head with its configuration and planar curves. Such unexpected understanding reflects an interesting interaction between the building users and the architects.

The relocated Zhanjiang Airport will serve the western Guangdong as a new gateway. The Western Guangdong is home to some unique cultures, including the primitive red soil culture which is originated from the geological characteristics of the local red soil, and includes traditional cultural elements such as bonfire dance, lion dance and dragon dance. So we also manage to incorporate

these local cultural characteristics into the terminal space.

The red symbolism of red soil culture is blended into the interior space. For example, the color elements of the seating in the terminal, and the red color visible from the ceiling gap all reflect the background color of local culture. The large space of the terminal features unitized suspended ceiling, each in diamond shape. With additional sound-absorbing perforation, each diamond is bisected into two triangles, adding another expression element. The diamonds and triangles form a sea of scales on the ceiling, echoing the oceanic characteristics of the region. As scales on both sides cascade, the gaps in between becomes larger, revealing and highlighting the red color as the background color of the local culture.

武汉火车站
Wuhan Railway Station, Wuhan

项目地点：武汉市，洪山区
设计时间：2006- 2008 年
建设时间：2006- 2009 年
建筑面积：106,841 平方米
合作单位：中铁第四勘察设计研究院集团有限公司，AREP
建筑师团队（广东省建筑设计研究院有限公司）：
陈雄、郭胜、周昶、邹泳文、巫仲强、盛晖、刘云强、黄咏梅
主要奖项：芝加哥雅典娜建筑设计博物馆颁发“国际建筑奖”
中国土木工程詹天佑奖
中国“百年百项杰出土木工程”
铁路优质工程勘察设计一等奖

Location: Hongshan District, Wuhan City
Design: 2006-2008
Construction: 2006-2009
GFA: 106,841m²
Partners: China Railway Siyuan Survey and Design Group Co., Ltd, AREP
GDAD Design Team: Chen Xiong, Guo Sheng, Zhou Chang, Zou Yongwen, Wu Zhongqiang, Sheng Hui, Liu Yunqiang, Huang Yongmei
Major Award(s): The Chicago Athenaeum Museum of Architecture and Design "International Architecture Award", Tien-yow Jeme Civil Engineering Prize, 100 Outstanding Civil Engineering Projects from 1900 to 2010 by China Civil Engineering Society, The First Prize of Excellent Railway Engineering Exploration and Design Award

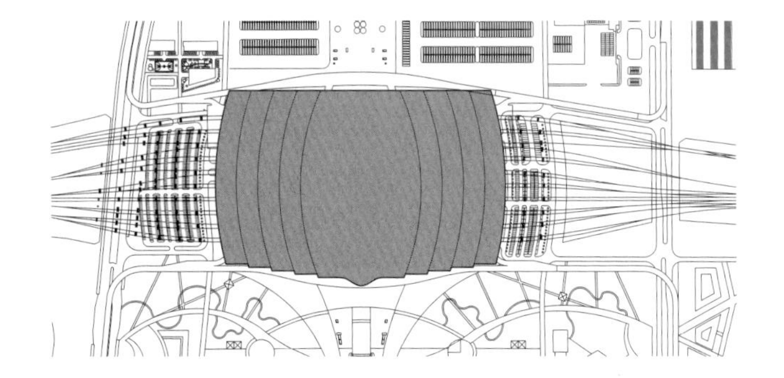

武汉火车站是中国第一条高铁线路——武广高铁的一个起始站，在中国高铁网络发展里程中具有比较重要的意义。由于武汉站是一个相对复杂的大型交通建筑，当时国内能够驾驭这样规模交通建筑的设计团队并不多，而我们刚好设计了广州白云机场 T1 航站楼，当时机场已经开始通航，所以武汉站的总承包——铁四院看中了我们在机场项目上的经验，便邀请一起去做武汉站的设计。

设计方的分工是这样，项目的外方设计单位是法国的法铁，他们做了方案设计和初步设计，我们团队的建筑专业参与施工图的设计，

结构设计方面由铁四院找了北京建研院的结构所负责。作为一个中外配合的项目，与广州白云机场 T1 航站楼采用了不完全一样的合作模式。

武汉站项目的创意是非常好的，建筑是一个复杂的形体，整个屋面系统与结构都比较复杂。为了实现这个创意，我们克服了不少困难，从最终的效果上来看，还是具有比较高的完成度，实现了最初的构思。这个方案有几个特色，首先是屋面的设计，屋面的天窗没有采用玻璃材质，而是选用了阳光板，这在当时是一种比较特别的天窗做法，效果更加清亮，可能受这个项目影响，现在很多火车站与高铁站的天窗都是用阳光板制作；另外，车道边的幕墙形式也比较有特色，在法方的设计概念下，我们也参与了幕墙系统的设计，包括幕墙的选材；还有一个特色，人们在进站的地方是可以看到轨道的，旅客一下高铁，也能看到顶上的高空间，突破了一些常规的模式。

武汉火车站毗邻国家 AAAA 级东湖风景区，周边环境十分独特，建成后的武汉站成为区域的交通中心、景观中心和城市新区发展的核心。

As the starting station of China's first high-speed railway line, Wuhan—Guangzhou high-speed railway, Wuhan Railway Station is of great significance in the development of China's high-speed railway network. Back then, most local design teams could be overwhelmed by such a large complex transportation building. Fortunately, GDAD was invited by China Railway Siyuan Survey and Design Group Co., Ltd, the General Contractor of Wuhan Railway Station, to join them for the project as they knew about our previous design experience with T1 of Guangzhou Baiyun International Airport, which was already put into use then.

AREP, the overseas architect office, completed SD and DD before GDAD's architect team proceeded with the CD design. BIAD Structural Institute retained by the General Contractor provided the structural design. The project joins the efforts and expertise of both local and international architects and engineers through a collaboration mode different from the one for T1 of Baiyun International Airport.

The project has a great creative idea while as a result, the building is complex in form, roof system and structure. We overcame many difficulties to realize the design idea. Fortunately, the final effect shows a high realization level of the original design.

The project has several highlights. First, its roof skylights use PP sheets instead of glass to achieve more transparent effect. PP sheet used to be a special solution, but has now been widely used for skylights at railway stations, probably because of the exemplary effect of this project. Second, the curbside facade is unique. We participated in the facade design, including material selection in support of AREP's design concept. Third, different from conventional design, tracks are visible at the departure area, and passengers can see the loft spaces overheads once getting off the train.

The project lies in close adjacency of the East Lake Scenic Area, a National AAAA Scenic Area, enjoying a superb environment. Upon its completion, the project has become a regional transportation center, landscape center, and the development core of new urban areas.

和谐号
CRH
和谐号
CRH

肇庆东站交通换乘枢纽

Zhaoqing East Railway Station Transportation Hub, Zhaoqing

项目地点：肇庆市，鼎湖区
设计时间：2016- 2017 年
建设时间：2016- 2018 年
建筑面积：85,000 平方米
建筑师团队：郭胜、陈雄、宋永普、罗志伟、罗文、林建康、黎智立、张栋、许秋滢、吴杰明、黄亦彬、敖立
主要奖项：行业优秀勘察设计奖 · 优秀（公共）建筑设计二等奖
中国威海国际建筑设计大奖赛优秀奖
广东省优秀工程勘察设计奖 · 公共建筑一等奖

Location: Dinghu District, Zhaoqing City
Design: 2016-2017
Construction: 2016-2018
GFA: 85,000m^2
Design Team: Guo Sheng, Chen Xiong, Song Yongpu, Luo Zhiwei, Luo Wen, Lin Jiankang, Li Zhili, Zhang Dong, Xu Qiuying, Wu Jieming, Huang Yibin, Ao Li
Major Award(s): The Second Prize of Excellent (Public) Architecture Design of National Excellent Exploration and Design Industry Award, Excellence Award of the Weihai International Architectural Design Grand Prix, The First Prize (Public Building) of Excellent Engineering Exploration and Design Award of Guangdong Province

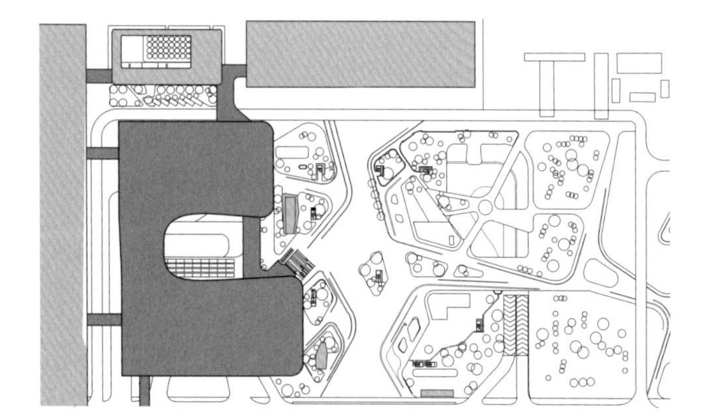

肇庆东站是支撑肇庆新区全面高质量发展的重要基础设施，跟肇庆新区体育中心是前后展开设计的项目。场地里有一个高铁站和一个城轨站，两条轨道十字交叉，均已通车运营，周边还有很多其他的交通方式，所以一个典型问题是怎么去缝合高铁、城轨和相关的交通，并实现这些交通和城市的整合。

首先，我们设计了一个换乘枢纽，将这些交通都缝合起来，在室内形成有序的转换，并在流线中增加商业配套，改善了换乘空间的品质。

将两种轨道交通的换乘从室外广场转移到了室内，再将大巴、出租车、私家车等其他交通方式按需要进行整合，在地下设置了一些中庭，改善负一层私家车的停靠环境，车道边也作一些提升。这个地块未来将通行地铁，我们也预留了地铁换乘的接口，将整体交通缝合进统一的流程中。

第二，对广场进行重新梳理，发挥了集散和景观的功能，为城市创造了一个很好的门户广场和城市客厅。相比于常规城市高铁站广场方正、空旷的做法，这个项目采用了比较自由的形态，加入了绿化景观，既考虑到大容量人流的集散，也兼顾了平时市民的休闲活动。场地设置一些绿化缓坡，缓坡下方设置了公共服务设施，将商业和洗手间等服务设施统一到景观设计中。

第三，我们在建筑本身还设计了一些灰空间，用作长途大巴的停靠功能，可以给出行的人们提供等候、休息的区域，也实现了遮阳避雨的功能，很典型地体现出岭南建筑人性化的空间特点。

As an important infrastructure to support the all-round and high-quality development of Zhaoqing New District, Zhaoqing East Railway Station was planned right after the Sports Center. The project site contains a high-speed railway station and an inter-city rail station, both under operation with two intersecting tracks. With other transportation modes around, the challenge is how to connect high-speed railway, inter-city rail and related transportation and integrate them into the urban fabric.

First, a transfer hub is planned to connect different transportation modes and ensure orderly indoor transfer. Commercial facilities are also embedded into the circulation to create better transfer spaces. The transfer between two rail transportation means is relocated from the outdoor square to the indoor space, and other transportation modes such as bus, taxi and private car are integrated as needed. Underground atria are provided to improve the parking environment of private cars on B1. Some improvements are also done at curbside. Accesses to the metro are reserved, so the transportation system becomes part of the process.

Second, the square design is reviewed to ensure its function for traffic distribution and landscape, and its positioning as an impressive gateway square and city parlor. Unlike those regular and open squares at the conventional high-speed railway stations, the square of the project is in free-flowing form with green landscape, which accommodates both the distribution of large number of passengers and the leisure activities of citizens. Some gentle green slopes, plus the public service facilities beneath, are planned at site to integrate commercial and toilet services into the landscape design.

Third, some grey spaces are planned in the building to berth long-distance coaches and provide sheltered waiting and lounge area for passengers, which reflects the typical people-oriented features of Lingnan architecture.

体育建筑

SPORTS BUILDINGS

体育建筑是功能性与标志性的集合体，具有运动的特征，从不同方面表达运动、动感、韵律感。

Sports buildings are both functional and iconic. With characteristics of sports, they reflect the senses of movement, motion and rhythm from different aspects.

广州亚运馆
Guangzhou Asian Games Gymnasium, Guangzhou

项目地点：广州市，番禺区
设计时间：2007- 2008 年
建设时间：2008- 2010 年
建筑面积：65,315 平方米
建筑师团队：潘勇、陈雄、王昵、刘志丹、宋定侃、邓载鹏、邵巧明、林建康
主要奖项：中国土木工程詹天佑奖
中国“百年百项杰出土木工程”
全国优秀工程勘察设计行业奖 · 建筑工程一等奖
中国建筑学会建筑创作优秀奖
AAA 亚洲建筑师协会荣誉奖
绿色建筑与低能耗建筑“双百”示范工程
詹天佑土木工程大奖 · 创新集体奖
中国室内设计年度优秀公共空间设计金堂奖

Location: Panyu District, Guangzhou City
Design: 2007-2008
Construction: 2008-2010
GFA: 65,315m²
Design Team: Pan Yong, Chen Xiong, Wang Ni, Liu Zhidan, Song Dingkan, Deng Zaipeng, Shao Qiaoming, Lin Jiankang
Major Award(s): Tien-yow Jeme Civil Engineering Prize, 100 Outstanding Civil Engineering Projects from 1900 to 2010 by China Civil Engineering Society, The First Prize (Construction Engineering) of National Excellent Engineering, Exploration and Design Industry Award , Excellence Award of the ASC Architectural Creation Award, ARCASIA Awards for Architecture (AAA) , One Hundred Green Building/Low-energy Building Demonstration Project, Tien-yow Jeme Civil Engineering Prize (Collective Innovation), Jintang Prize-China Interior Design Award

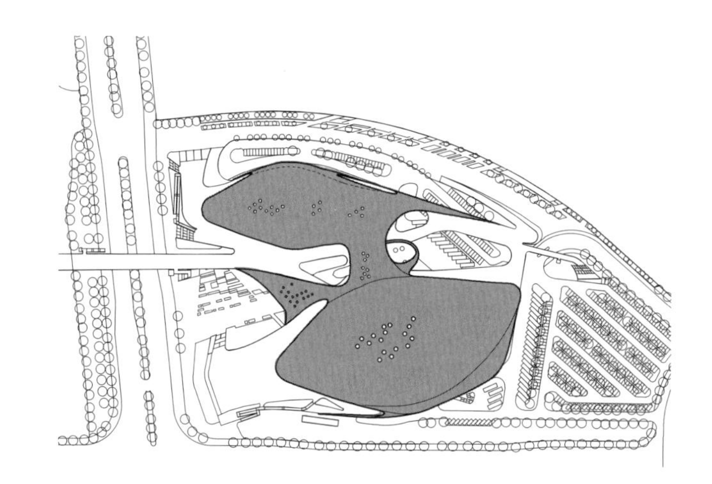

广州亚运会主场馆这个项目是机场所 ADG 团队意义重大的一个项目，也是广东省建院继广州白云机场 T1 航站楼之后一个非常重要的项目。T1 航站楼是 2004 年启用的，而这个项目实际上是 2007 年年底启动国际招标，到 2008 年中标。在 T1 航站楼积累了做大型复杂公共建筑的经验和技术，在这个项目实践了这些复杂项目的经验。

这个项目的意义是在 T1 的基础上，以独立的身份去参加当时广州市组织的高规格大型国际竞赛。当时为了亚运会，这是新建的唯一主场馆。广州市组织了这次非常高规格国际竞赛，来的包括国内外一些著名的设计机构、事务所和公司，最终选出了 8 个团队。 8 个团队里，有 6 个是国外的团队或者是中外合作的，只有两个是独立身份，ADG 团队是其中一个。最终这个项目是中国建筑师在重大项目的国际竞赛中，原创中标并高完程度实施的案例。这在当时以及直到现在也是比较少见的。这个项目对于中国建筑师群体，对于广东省建院，对于机场所，对于 ADG 团队都具有非常重要的里程碑意义。

这个项目为什么能胜出？项目在广州亚运城，选址其实并不大，要建的功能很多，包括一个体操馆、一个台球馆、一个壁球馆，还有一个亚运的博物馆，总共相当于四个馆的功能，建在这块并不是很大的场地上。建筑面积是 6 万平方米，然而这块地本身不大，却要解决这么多功能，还要考虑停车、景观绿化和集散广场，另外要和地铁站衔接，同时周边还有一个项目也是这次投标一起做的——亚运媒体中心，是连在一起的。

从地铁站到媒体中心再到广州亚运馆，是一条线，由一个高架的二层平台连起来，整个是一体化的规划设计，第一次创造性地把体育馆群用一个统一的屋面，把整个建筑形体组合在一起，而不是分场分散各自独立。在这之前无论是国内还是国际上，绝大部分的体育场馆都是分离的，尤其是国内，当时都是典型的体育中心——一个体育场、一个游泳馆、一个体育馆分开的，像天河体育中心那样的分离模式。在这么一个用地紧张的情况下，如果按传统的模式去做，其实很难做到比较好的环境和功能平衡。而这是第一次创造性地把

它们连在一起，产生了一些积极开敞的灰空间，使它变成一个群体，在集约的用地下创造了非常好的、高品质的公共空间，这些公共空间面向赛时和赛后的利用。赛时是集散的地方，有一些室外广场，有一些室外的灰空间；赛后这些空间又提供给市民高品质的开放空间。后来有机会跟一些运营方聊起，他们说除了馆里面的空间有使用价值，这些室外和半室外的灰空间其实也蛮有价值，一些活动可以在灰空间里进行，体验很好。

还有就是这里面有一系列比较好的创新，包括结构体系，在这样的大跨度结构体系上面，采用双向的网络结构，还能够做出比较干净的室内空间界面，这是一点。

另外一点，也是这个项目很重要的一点，是复杂非线性三维的一个高完成度的实现，这是国内很早期用犀牛软件去实现复杂非线性三维曲面的一个项目。用犀牛软件建构了建筑的表皮和建筑的形态，之后结构专业也用犀牛软件建构结构，包括整个表皮。后来施工的所有数据也是根据这样的建模数据去进行加工和安装。这是我国早期实现复杂非线性三维的一个建筑，这个屋面系统在当时业界是非常领先的。

幕墙系统是由水平横梁、抗风构件、遮阳构件和垂直不锈钢拉手组成的，是很通透的一种幕墙形式。幕墙的结构是跟主体的混凝土柱连在一起。所有的柱子，前面都有个“X”形结构延伸出去支撑水平的横梁，没有额外的二次结构。后来广州白云机场 T3 航站楼也是用了这种一体化设计的方式。T3 航站楼因为柱子直接靠着幕墙，所以它也是用这样的方式，但是 T3 的柱网更大，为 18 米，广州亚运会场馆大概在 12 米，两个项目柱网大小不同，因为形体相对比较复杂。所以这个设计整体上比较丰富，属于比较简约的一个体系，是幕墙跟主体结构一体化的设计。

另外还有一个很重要的难点，是亚运博物馆这个体形。它是一个大悬臂，大概悬臂 20 多米的红色形体。要实现这个形体，结构是比较

复杂的，建筑跟结构要有一个很好的配合。我们是用交通核作为一个主要的支撑，整个结构再用一个大拉杆把它拉住，然后再悬挑出来，拉杆藏在结构的屋面体系上，相当于用筒体把它拉住以后，它才悬挑出来那么多。这个筒体本身也要有一个相对比较大一点的基础才能稳，所以这个结构有一个比较大的创新点，也是建筑跟结构一起配合实现出来的一种大悬挑形体，在这个领域具有一定的创新性。

我们对参观流线进行了重点组织，比如人从交通核一直坐电梯上到观光厅，再从螺旋坡道一直走下来，边走边观赏亚运的展览，类似于古根海姆博物馆，但在场地里内与外形上是看不出这个特征。所以整个设计其实是非常有创意的，还刊登在《建筑学报》的封面，当时我们去投稿《建筑学报》，大家一看觉得这确实很棒，而且是关于亚运会的。这个项目也是在 2008 年奥运会之后，独立原创设计中标实施的案例。

项目的造型主题是自由体操的彩带舞，用飘逸彩带这个概念去做这个项目，形体像彩带舞形成的流畅曲线。从人的视角看，是一个多面的连续曲线变化的造型，所以其实它是一个没有灭点，同时非常流畅的形体。人从不同的视角去看，它都是一个完整而又不同的形态，是很完整的体验。它的艺术感很强，结合自由体操的主题，反映的是体育建筑那种柔美而非力量感的线条，给人流畅、灵动的感觉，不只是有力量感的体育建筑，这个也是体育建筑形态上的创新。

The project is of great significance not only for ADG, but also for GDAD after we completed the design of T1, Guangzhou Baiyun Airport. The expertise and experience we gained from T1, a gigantic complex, have been well applied to the project.

The significance of the project for us lies in the fact that ADG, as an independent participant, participated in an influential high-standard international competition in Guangzhou, together with some well-known design offices at home and abroad. Eight teams were selected as finalists, of which six were international teams or China-foreign consortia, and ADG was one of only two teams that competed in an independent capacity. Eventually, we won the competition with our originally created design proposal and realized the project to a high degree of design. This was one of the few projects where a Chinese design institute won a competition and realized the project with independently created design proposal after the Beijing Olympic Games in 2008. It marked a milestone for Chinese architects, GDAD, the Airport Office and the ADG team.

Located in Guangzhou Asian Games City, the project is planned with various functions on a limited site, including a gymnasium, a billiards hall, a squash hall, and a museum for the Asian Games. With a floor area of 60,000 sqm, the project has to accommodate parking, landscaping and distribution square , and provides a connection with the surrounding Asian Games Media Center and metro station.

In the design, we creatively brought all building blocks under the same roof, while before that, most stadiums and gymnasiums at home and abroad had been in a separated layout, such as Tianhe Sports Center in Guangzhou. Given the limited site area, a conventional solution would not possibly achieve a good balance between environment and functions. So we connected buildings together and created some positive, open and quality grey spaces. These spaces can be used for spectators distribution during the games, and serve the public after the games. Later when I happened to talk to someone working with the operator. They said that in addition to the space inside the gymnasium, these outdoor and semi-outdoor gray spaces have been quite valuable, as they can accommodate some activities and bring very good experience to people.

The project also implemented some very good innovations. First, despite such a long-span structure system, a two-way network structure is employed and a relatively neat indoor space interface is created. Second, we achieve a complex nonlinear 3D shape in high conformity with the design, using Rhinoceros to construct the building skin, form and structure. In construction, fabrication and installation are based on the modeling data. So it is an early example of complex nonlinear 3D building realized in China. The roof system was at a leading position in the industry then. Third, the curtain wall system is composed of horizontal beams, wind-resistant members, sun-shading members and vertical stainless steel handles. It is very transparent and realizes connection with the main concrete columns. All columns have an X-shaped structure extending outward as horizontal beams. The design adopts a relatively concise system which integrates curtain walls and the main structure. Another major challenge is the building form of the Asian Games Museum, a red block with over 20 m cantilever. Considering its complicated structure, we use the traffic core as the main support, and then a big pull rod to hold it, which is a creative solution.

We put emphasis on tour routes. Visitors take the elevators all the way from the traffic core to the observation hall, and watch the exhibition of the Asian Games while walking down the spiral ramp, similar to the visitor experience in the Guggenheim Museum. This creative design was featured on the front cover of the Journal of Architecture.

Conceptualized as flying ribbons, the project presents a smoothly flowing curved form without vanishing point when viewed from human perspective. This highly artistic form echoes the theme of gymnastics, while the gently undulating lines of the buildings portray a free-flowing and lively sports building different from those emphasizing power and tension, hence an innovation in building form.

116、117

汕头亚青会场馆
Shantou AYG Gymnasium, Shantou

项目地点：汕头市，澄海区
设计时间：2019- 2019 年
建设时间：2019- 2021 年
建筑面积：146,400 平方米
建筑师团队：潘勇、陈雄、庞熙镇、易田、罗文、戴志辉、黄亦彬、孙家明、邵国伟、黄河清、袁峻豪、梁蕙茵、谢海彬

Location: Chenghai District, Shantou City
Design: 2019-2019
Construction: 2019-2021
GFA: 146,400m^2
Design Team: Pan Yong, Chen Xiong, Pang Xizhen, Yi Tian, Luo Wen, Dai Zhihui, Huang Yibin, Sun Jiaming, Shao Guowei, Huang Heqing, Yuan Junhao, Liang Huiyin, Xie Haibin

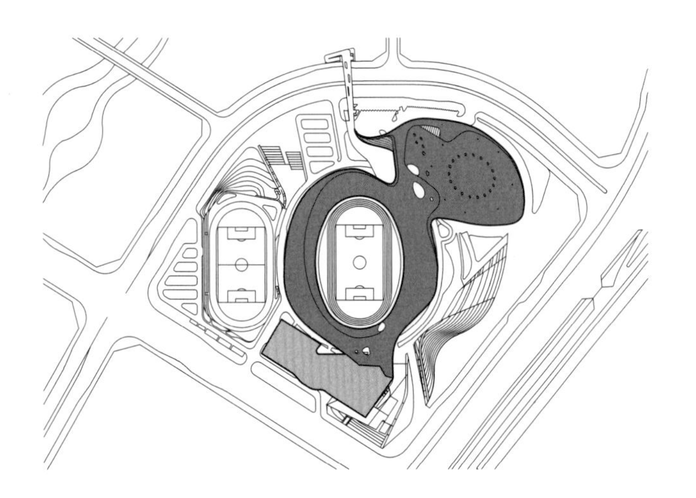

汕头亚青会场馆是继广州亚运会场馆设计 10 年之后，团队做的另外一个国际赛事场馆，是亚洲青年运动会国际赛事的一个主场馆。整个设计也是用集约的布局去解决用地紧张的问题，同时也创造了很多比较好的灰空间，作为面向城市生活的一些公共空间。除了体育功能，这些公共空间也是作为赛时赛后的利用。可以看到 10 年之后，有些基本的思想还是一脉相承，但是增加了一个会议中心的功能，同时因为在海边，与海洋的元素和与滨海城市的一些特质更能进行呼应。跟广州亚运会场馆不同的是，因为有体育馆、体育场，还有会议中心这三大功能在里面，功能性便更强大。包括开闭幕式的策划和点火的路径，这些都在设计里进行了考虑。最终整个项目的完成度很高，也受到各方的高度评价。

项目的选址与汕头大学的一个新校区结合起来，希望能够在赛后除

了为城市服务，还可以为大学服务，这是一个想法。另外一个想法是通过一个高架把滨海的景观带连在一起，从会议中心到海边，再到学校这么一个路径。同时它还结合会议中心，创造了一系列公共活动的空间。

屋面上也创造了一系列公共活动空间，可供悠闲看海、休憩，设计了一些可以停留的平台，作为市民休闲交往的公共空间，非常有意思。它的结构是用了一个轮辐式作为结构体系，这是它的一个特色。丰富的建筑造型与空间，很重要的一点就是三维的建构，利用犀牛软件进行整体统一，包括建筑、结构、幕墙和金属屋面的建构，是多专业交叉融合的体系。

项目的分板衔接也做得非常到位，从设计的推敲开始，因为不同的面要做不同的分板，而这些面之间的一些分板技术，在 3D 模型上就推敲好了，直接指导现场安装，达到了很高的完成度。很多复杂形体或者非线性三维形体，如果分板做不好的话，其实收边收口也是没法做好。这个空间复杂的形体，对分板的编织交接进行了很好的推敲，这是很长时间推敲的结果，甚至有些地方是每块板去推敲，才能把它衔接好。这些复杂非线性三维曲面的组合，我们不仅看它整个面的形体大感觉，同时要细看这些分板，这就是精细化的设计水平。

项目有很高的完成度，这也确实是在亚运馆基础上的设计，比亚运馆更为丰富，更有一种魔幻的感觉。这个建筑的语言是很流畅的，用了很前卫的建筑语言去表达。而且，我们是独立、原创的团队，这个也是非常难得的。

We were engaged to design the AYG Stadium in Shantou, another international competition venue, ten years after the design of Guangzhou Asian Games Gymnasium. For this project, a compact layout is proposed to address the tight site area; meanwhile, various quality grey spaces are created as public spaces for urban life. Apart from sports functions, these public spaces can also be used during and after the Games. This shows an extension of our design ideas from the Asian Games Gymnasium ten years ago, though a conference center is added to AYG Stadium. In consideration of the coastal location of the project, the design also responds to the marine elements and characteristics of coastal city. Different from Guangzhou Asian Games, the project contains more functions with three major components, i.e. gymnasium, stadium and conference center. The design also considers factors like the opening and closing ceremonies and the path to light AYG flame. Eventually the project was realized in high conformity with the design and acclaimed by from all parties.

The project site is combined with a new campus of Shantou University, so that the project will serve both the city and the university after the games. Besides, an elevated road is planned to create a coastal landscape belt from the convention center to the seaside, and further to the university, and to provide a series of public activity spaces based on the conference center.

Public activity spaces are also created on the roof where one may enjoy sea view and take a rest. Some platforms are designed as fun public spaces for citizens' leisure and and communication. Structurally a spoke-type system is employed as a highlight of the project. To provide diverse building forms and spaces, 3D construction serves as important support. Rhinoceros software is used to integrate the buildings, structures, curtain walls and metal roofs into a multi-disciplinary system.

Facade panel division and connection are well handled. As different facade surfaces require different panels, the approach to divide panels between different surfaces is discussed and finalized on 3D model, so the panel installation can be conducted right at site in high conformity with the

design. For complex or nonlinear 3D blocks, it's impossible to realize nice edges or ends without proper panel division. For this project with complex spaces, the panel weaving and connection are well studied for a long time; sometimes, even each panel has to be studied to realize a satisfactory outcome. The combination of these complex nonlinear 3D curved surfaces not only appears generous on the whole, but also delivers elaborate panel details that showcase the refined design expertise.

The project is completed in high conformity with the design based on our experience with the Asian Games Gymnasium. Yet it appears more diverse and magical than the latter thanks to the coherent and even avant garde architectural vocabulary. Also it's not so common for a local design institute like us to independently deliver the original creative design for this important project.

广州花都东风体育馆
Dongfeng Gymnasium, Huadu District, Guangzhou

项目地点：广州市，花都区
设计时间：2008 年
建设时间：2008- 2010 年
建筑面积：31,416 平方米
建筑师团队：郭胜、陈雄、陈超敏、陈应书、邓弼敏、黄蕴
主要奖项：全国优秀工程勘察设计行业奖 · 建筑工程三等奖
中国室内设计大奖赛学会奖
广东省优秀工程设计二等奖

Location: Huadu District, Guangzhou City
Design: 2008
Construction: 2008-2010
GFA: 31,416m^2
Design Team: Guo Sheng, Chen Xiong, Chen Chaomin, Chen Yingshu, Deng Bimin, Huang Yun
Major Award(s): The Third Prize (Construction Engineering) of National Excellent Engineering Exploration and Design Industry Award, China Institute of Interior Design (CIID) China Interior Design Awards, The Second Prize of Excellent Engineering Design Award of Guangdong Province

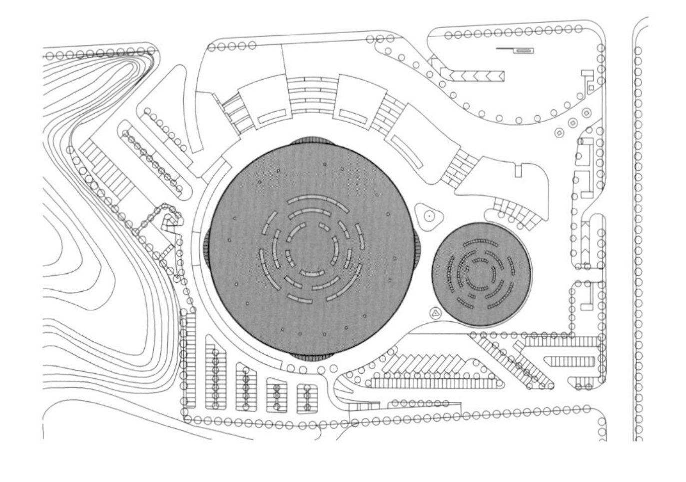

广州花都体育馆是 2008 年国内公开竞赛中标的项目。项目主要功能是一个篮球馆，当时定位为亚运会的备用场馆，选址在广州花都区，周边是汽车城，有一些产业与居民区，因此希望赛后能够为周边区域的居民提供一个比较好的体育设施。

基地是一个坡地，选址在一个公园的旁边，有一些小山头，在这样一个坡地下面，怎样去有效地解决好功能的布局？设计用了两个不同大小的圆的方式，用平台把它们连起来，同时结合旁边公园的山体，立体地解决交通。人行通道位于平台上方，车行位于平台下方，人车分流，解决了山地的交通组织。有这些平台之间的衔接，又有一些灰空间在里面，这是非常有意思的。

建筑的形体是类似圆的球体，构成上看，是非线性三维形体，里面是比较规整的一个篮球馆，圆形体育馆具有其合理性，即使它是圆

形，比赛的场地还是长方形，主要的两个观赏面座位会多一些，观看的位置正好可以充分利用圆的合理性，是一种典型的体育馆类型，符合好视线的座位区域比较多的布局。

项目的结构跟建筑之间的配合也是非常到位的，是很一体化的一个设计，整个结构相当于两个圈，一个是顶上的一个圈，另外一个就是在观众席后面由一个“人”字形结构支撑的一个圈。两个圈之间用桁架撑起来，是跟整个造型很贴合的设计。上下两个圈利用径向一榀一榀的桁架撑起来，形成立体空间。下面的圈跟观众席下方的座位不是用直的柱子支撑的，而是“人”字形的柱子支撑。可以这么说，这是延续了广州白云机场 T1 航站楼的一些特征，“人”字形结构的做法，整个建筑的语言是流畅的，这样上面和下面的圈的连接是非常有韵律的。

要设计这样的体育场馆空间，结构跟建筑要配合得很好，通常的体育场馆是不吊顶的。因此建筑和结构之间的配合要很到位，不像交通建筑，常规还是会做一些吊顶。可以看到这样一些结构，包括环形的马道，跟体量连在一起。

关于天窗的设计，在外面是可以看到天窗的，我们希望从整个外围来看，天窗是一种像素化的设计，怎么与室内过渡是个问题，因为室内若也像素化，看起来效果就可能会比较乱。实际上屋面的像素化天窗，有一部分天窗玻璃下面是有窗的，有一部分是没窗的。这样和室内过渡，使得室内看 8 个天窗是整齐的，外形看它是一种充满动感的玻璃图案，是一个很有意思的设计，完成度也比较高。当然跟外壳的分板也有关系，因为它是由中间那个点往外分，一层一层，有些板是到一定程度再分一层，多层过渡，用合理的分板尺寸，去实现这么一个类似球面的表皮。

We won the design contract through an open design competition in 2008. The project with a basketball hall was planned as a backup facility for the 2010 Asian Games. Given its location adjacent to the automobile city, some enterprises and residential compounds, the project was also expected to provide good sports facilities for the residents after the Games.

The site is sloped besides a park and some hills, making it challenging to create an effective functional layout. As response, two circles in different sizes are designed and connected via platforms, which, together with hills by the park, ensures circulation of the project. With footpaths on the platforms and vehicular lanes below, pedestrian are separated from vehicular circulations to realize a satisfactory traffic organization in the hilly area. It is also fantastic that the platforms are connected and some grey spaces are created.

The building appears almost like a round sphere, while structurally it is a nonlinear 3D block with a relatively regular basketball hall inside. A round gymnasium makes sense, as the competition field would remain rectangular anyway, and more seats are possible along the two main sides with good vision. This is a typical gymnasium design.

The project features a highly integrated design of structure and architecture. The whole structure consists of two circles, one at the top and the other supported by a herringbone structure behind the spectator stands. The two circles are supported by radial trusses, creating a 3D space and fitting well the whole building form. The lower circle and the seats under the stand are supported by herringbone columns instead of straight ones. This carries forward some features of T1 of the Baiyun International Airport in Guangzhou. The herringbone structure contributes to coherent

architectural language, and provides a rhythmical connection between the upper and lower circles. Unlike transportation buildings, a stadium usually doesn't have suspended ceilings. Creating such a stadium space requires a good coordination between structure and architecture. So some structures, including the circular catwalk, are connected with the building volume.

Regarding skylights, they are visible from the outside and pixelated. The problem is how to realize the transition into the interior spaces, because a pixelated interior would be a mess.To tackle this problem, pixelated skylights are made partially windowed, so that a good transition to the interior space is achieved; meanwhile, the eight skylights look neat from indoors, and form a dynamic glass pattern in building appearance. This is a very interesting design and successfully realized. Of course, this is also related to the panel division of the building skin.Specifically panels are divided from the middle point to the outside, layer by layer, and some are subdivided when reaching a certain point. Reasonable size of panel division help realize such a spherical surface.

肇庆新区体育中心
Zhaoqing New Area Sports Center, Zhaoqing

项目地点：肇庆市，鼎湖区
设计时间：2015 年
建设时间：2016- 2018 年
建筑面积：85,694 平方米
建筑师团队：郭胜、陈雄、陈超敏、罗志伟、宋永普、林建康、陈艺然、罗文、黄亦彬、黎昌荣、周志明、敖立、何业宏
主要奖项：行业优秀勘察设计奖 · 优秀（公共）建筑设计一等奖
中国建筑学会建筑设计奖 · 公共建筑三等奖
中国威海国际建筑设计大奖赛铜奖

Location: Dinghu District, Zhaoqing City
Design: 2015
Construction: 2016-2018
GFA: 85,694m^2
Design Team: Guo Sheng, Chen Xiong, Chen Chaomin, Luo Zhiwei, Song Yongpu, Lin Jiankang, Chen Yiran, Luo Wen, Huang Yibin, Li Changrong, Zhou Zhiming, Ao Li, He Yehong
Major Award(s): The First Prize of Excellent (Public) Architecture Design of National, Excellent Exploration and Design Industry Award, The Third Prize (Public Building) of ASC Architectural Design Award, Bronze Award of the Weihai International Architectural Design Grand Prix

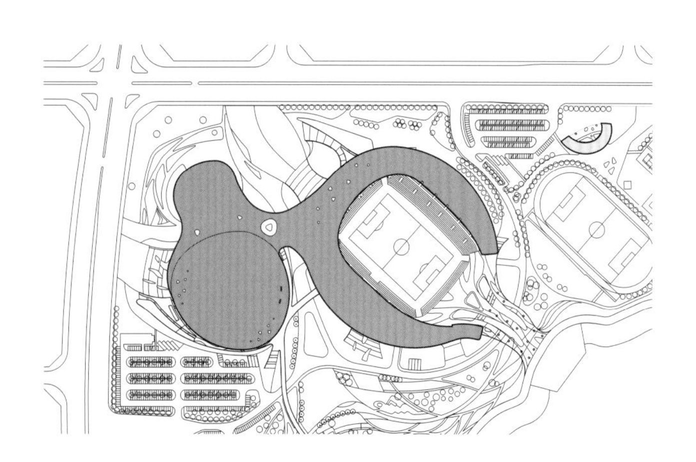

肇庆新区体育中心项目在我们投标的时候是位于新区的另外一块地上，这块地当时有体育场、体育馆、游泳馆和训练馆等，是占地比较大、功能很齐全的体育中心规划。但是投标中标之后，换了一块用地，这是政府经过深入的思考，认为使用土地和功能更加优化配置的结果。

当时决定把肇庆市旧的体育中心进行改造利用，新区这块用地的规模缩小了，把功能更聚焦在足球的主题上。这样相当于将旧的部分提升，将新的部分聚焦专业化，大大减少用地资源，也满足省运会的功能，是可持续发展的思路。

在集约用地下，我们再次采用了集群式的手法去设计，其实回过头

看这已成为我们团队一种标志性的体育建筑策略。这个项目是由体育馆、训练馆以及作为专业足球场的体育场，组成一个整体去做的设计，是连成一个整体的非线性曲面。项目同时是赛时利用和赛后整个城市的市民中心，是以公共活动为目的来创造的高品质空间，整个形体是很灵动的，连在一起产生了一些非常有意思的灰空间。

项目位于新区长利涌的自然悬河旁边，那是作为泄洪功能的一条涌，自然的形态是很灵动的。在旁边的这块地我们呼应这样的自然环境，包括肇庆很优美的自然山水，所以建筑的形态非常灵动。同时在悬河高出地面大概五六米的地方，围绕悬河周边有一个城市的公共景观带，景观带正好跟体育场 5 米到 6 米的观众层连在一起，是非常有意思的联动。

另外还有一个创造性的部分，是把体育场打开一个缺口，汕头亚青会主场馆也是这样，打开缺口后能够看到大海。这个地方打开的缺口是朝南的，打开后就看到长利涌，是一处很好的景观。作为体育中心主要是以东西看台为主，那么我们打开南面是可以的。南面其实在平台下区还是有座椅的，是一个环状的坐椅区，形成了比较好的氛围。在架空层上面那一区段，打开以后面向长利涌，有比较好的室外景观。所以举办体育赛事的时候，有自然借景的想法在其中，包括开闭幕式，有一些场景，都可以通过这个地方实现比较特别的场景营造。

这个项目集约了用地之后，剩下的一些场地主要是作为专业的足球场，取消了跑道，取消后整个看台跟足球场之间的距离是很近的。

这样的话对于比赛的氛围和座椅区视线的质量都特别好，这是专业性的条件。

因为是体育建筑，建筑与结构和机电之间的配合是需要非常到位的，特别是跟结构之间的配合。从这个体育馆的角度来看，它是轮辐式的结构，就像一个轮胎放在上面。还有一些拉杆，整个拉起来是很轻量化的结构，也很通透。因为是拉杆结构，基本上感觉不到结构的体量。而且我们把它喷上深色以后，更使它弱化了结构的体量感，使场地和室内的体验更加舒适。另外一点是这个体育场是用了三角形的箱型梁，通过两层的支柱支撑，作为整个悬挑的体育场罩棚。这样的设计建筑和结构配合得很好，同时也把结构作为建筑装饰的一个部分和一个构件来表现。

另外还有一个比较精彩之处，就是在灰空间里面有一个花篮状的结构，是很经典的设计，这是建筑与结构配合设计的灰空间花篮状支撑结构，是结构逻辑和建筑美学逻辑高度结合的一体化结构。作为灰空间画龙点睛之处，有比较高的业界评价。

肇庆是一个山水城市，同时是很出名的砚都，所以当时参加投标的团队做了各种砚台组合的概念，都是来自于砚台之类的灵感。后来换了新的地块，我们把之前的概念抛弃了，最后把砚生水墨这个概念放在里面。我们想砚本身是文具，但是它的精华之处是水墨。水墨，可以画画，可以写书法，我们希望把砚生水墨这个概念用到很流畅的形体组织里面，没有再说用砚台的概念去做，得到了业主的赞赏。

At the bidding stage, the project was planned on another plot in the New District as a large sports center with fully-fledged functions, including a stadium, a gymnasium, swimming pools and training halls.However, after we won the bid, the government decided to develop the project at the existing site for a better match between the land and the functionality.

The new plan is to redevelop the old sports center in downtown Zhaoqing, and use the new site mainly for football facilities. This is more sustainable as it can largely cut the land use and meet the needs of the provincial games. Following the principle of intensive land use, we once again adopt cluster design to join all sports buildings and form a holistic and nonlinear surface. The project is planned as both a competition field during the games and an urban civic center afterwards. It provides high-quality spaces for public activities, and the lively forms create some interesting grey spaces when connected.

The lively building form is designed to respond to a nearby natural perched river called Changli Canal and beautiful landscapes in the distance. An urban public landscape belt around the perched river is connected with the stand floor between 5 m and 6 m of the stadium, hence an interesting interaction .

To bring natural scenery into the project, the south of the stadium is left open. In this way, Changli Canal is visible from inside the stadium, and a highly unique scenario can be created during the games.

With a land-intensive concept, the site is mainly used as a football pitch. The runway is canceled to bring stands very close to the football pitch and realize the required competition conditions.

A sports building requires good coordination between architecture, structure and MEP. This gymnasium has a spoke structure, as if supporting a tire. The tie rods, when pulled up, appear very light-weighted and transparent, almost dissolving the sense of structural volume. After they are painted dark, the sense of structural volume is further reduced, bringing about more comfortable indoor experience. Besides, the triangular box beams are supported by two-storied pillars, serving as a cantilevered canopy over the whole stadium. Through close collaboration of architects and structural engineers, structural system is realized as a part and component of building decoration. Another design highlight is a flower basket structure in the gray space, which is quite classic. It is an integrated structure reflecting a high combination of structural logic and architectural aesthetic logic. As a finishing touch of the grey space, it is highly acclaimed in the industry.

In addition to the picturesque mountains and waters, Zhaoqing is also known for inkstones. So a few inkstone options were developed and eventually the one reflecting how inkstone produces ink was incorporated into the design. For us, inkstone is a stationery, while its true meaning is about making Chinese ink which can be used for painting and calligraphy. This concept helped create a free flowing building form and was well accepted by the client.

顺德德胜体育中心

Shunde Desheng Sports Center, Foshan

项目地点：佛山市，顺德区
设计时间：2018－2020 年
建设时间：2020 年至今
建筑面积：210,709 平方米
建筑师团队：郭胜、陈雄、陈超敏、刘德华、林建康、黎昌荣、邓载鹏、陈伟根、翟卓越、江祖平、伍思迪、韦锡艳、莫颖媚、黄玉青、凌浩智
主要奖项：中国威海国际建筑设计大奖赛优秀奖
广东省注册建筑师协会广东省建筑设计奖·建筑方案奖公建类二等奖

Location: Shunde District, Foshan City
Design: 2018-2020
Construction: 2020 to present
GFA: 210,709m^2
Design Team: Guo Sheng, Chen Xiong, Chen Chaomin, Liu Dehua, Lin Jiankang, Li Changrong, Deng Zaipeng, Chen Weigen, Zhai Zhuoyue, Jiang Zuping, Wu Sidi, Wei Xiyan, Mo Yingmei, Huang Yuqing, Ling Haozhi
Major Award(s): Excellence Award of the Weihai International Architectural Design Grand Prix, The Second Prize (Architectural Design - Public Building) of the Guangdong Architectural Design Award of Guangdong Province Registered Architect Association

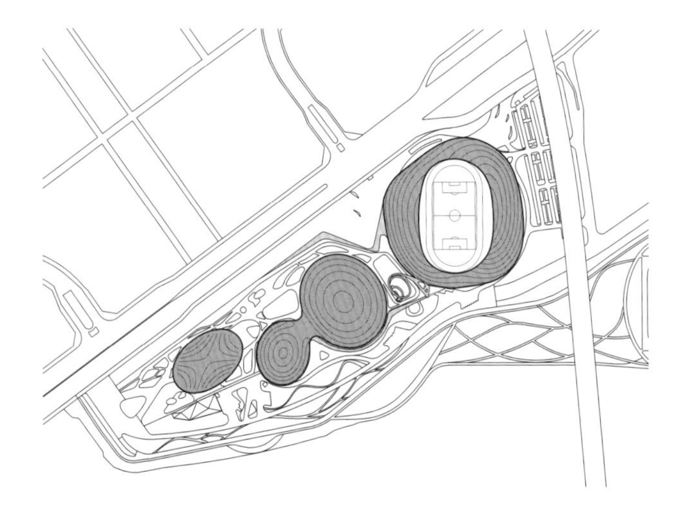

顺德德胜体育中心是我们一个重要的体育建筑作品，它是 2018 年全国竞赛中标的，我们再次用了集约的布局方式去解决城市高密度空间的平衡问题。

整块用地分为两部分，上面整个高架是广州南站通往珠海的城轨。城轨高架把用地分成了两块，另外一块用地旁边远期有地铁站，所以靠近那一块用地是可以做 TOD 开发的。

当时我们经过不同布局的比选，考虑到有高架路，虽然下面是连通的，但是在现场看，这个用地明显是分成两块的。那么我们怎么去解决这个问题呢？如果说把体育场放在 TOD 附近那块地，相对来讲是比较宽松一点。但是那块地不能放两个馆，如果放了体育场在那边，这边还有训练馆和游泳馆，这样的话势必可以放得下，但是整个形

态会很离散，被中间一条高架轨道一分为二，这种效果也不好。后来我们反复比选，是所有参赛单位里唯一一个把所有四个主要场馆集中在这一块地上，把另外一块有 TOD 开发价值的地留给了城市。这样整个形态是非常统一的，管理也非常好，同时还多了一块地作为 TOD 开发，可以跟体育场互动，形成互生的关系，这个也是我们最大的一个亮点。我们经过反复的测算，在类似长条三角形的一块地上精心地布局，正好把体育场沿着长轴布置下去，这是关键问题，把体育场的问题解决了，其他部分就可以排下来了。这是体育场和体育馆集群的一个设计，用了这样的一种方式，更符合解决紧凑用地公共空间的品质问题。

除了首层和 5.5 米层，还用一块大斜板把几个场馆的屋面连在一起，去满足这种相对高密度的体育场馆群疏散及活动空间的要求，这是一个立体开发的模式，是国内首创的。它仍然有传统的首层来解决贵宾和运动员进场的需求，而 5.5 米层是解决普通观众入场的需求。

斜向的屋面层把公共空间串在一起，为城市创造了非常丰富的空间，整个屋顶和景观花园一样，跟下方楼层又形成立体丰富的灰空间。在灰空间中穿插了商业设施的布置，使它对于城市来说有一个非常友好的界面。这一连串的商业设施，同时又为体育场馆赛后利用创造了很好的空间格局。

从屋面下来的路径，也做了精心的设计。螺旋的坡道让人从地面直接往返屋面，形成了立体的环形流线。德胜河是一个比较流畅的河道，我们希望项目的整个形态是灵动的，跟水的文化能够协调，并且有机地结合起来，把地方特色和岭南建筑在自然环境的营造中结合起来。

场馆追求轻盈，质轻而柔的建筑形象，采用玻璃幕墙、铝板与索膜材料的结合。体育场采用巨型框架格构柱与金属屋面的结合，打造立体流畅的建筑形象；体育馆、训练馆采用Ⅴ形柱与曲面顶棚结合，形成一个流畅优雅的外壳；游泳馆采用索膜材料，营造轻盈通透的效果；四个场馆统一于立体绿化屋面之下，为满足大跨度要求，绿化屋面采用双向平面桁架，并在主桁架设置预应力拉索，以控制竖向变形，楼板采用钢筋桁架楼承板。通过建筑结构的密切配合，打造整体和谐统一又极具标志性的建筑形象。

We won the national competition for the project in 2018. In this important sports building project, we again adopt a compact layout to balance against the high-density urban spaces.
The site is divided into two parts by the elevated intercity rail line from Guangzhou South Railway Station to Zhuhai. A metro station planned nearby for the far future makes TOD possible on plot near the station.

While comparing the different layout options, we realize that, though the site is continuous at grade level, it is virtually divided by the elevated railway overhead. If we place only the stadium on the TOP plot, the plot would appear spacious; but we have to place the training hall and the natatorium on the other plot, leading to fragmented layout and poor connection between the venues. After rounds of studies, we decided to place all four major venues on one plot and leave the other with TOD potential to the city, and we were the only bidder to do so. In this way, the site has a coherent layout, the management becomes easier and more importantly, one plot is spared for TOD to interact with the stadium. This is the biggest highlight of our design. Though careful calculation, we were able to place the stadium along the long axis of this long triangular plot. After this key issue is properly addressed, the other buildings can be readily arranged. The clustered design of stadium and halls better accommodate the quality issue of public spaces in a compact land area.

Apart from the F1 and 5.5m level, a large sloping board is also adopted to connect the roofs of the several venues. This multi-level development mode is the first case of its kind in China and can well address the needs of evacuation and activity space in high-density sports venues. Meanwhile, a typical F1 is still provided to accommodate the entry of VIPs and athletes, while level 5.5 m handles the entry of ordinary spectators.

The inclined roof links public spaces together, creating varied spaces for the city. Together with the landscape garden, it forms diverse grey spaces on different levels. Some commercial facilities are provided here and there in these gray spaces, offering a highly friendly interface for the city and creating a favorable spatial pattern for the post-game use of stadiums.

The path down from the roof is also carefully designed. The spiral ramp serves as a multi-level circular circulation and enables a return trip directly from the ground to the roof. With reference to the gently flowing Desheng Canal nearby, we hope the project also presents the free-flowing form and echo with the water elements; moreover, blend the local features with Lingnan architecture in such a natural context.

To portray a light-weighted and gently undulating form of the venues, the facade features a combination of glass curtain wall, aluminum panel and cable-membrane structure. Specifically, the stadium combines giant frame lattice column with the metal roof to create a free-flowing building image; the gymnasium and training hall adopt V-shaped columns and the curved ceiling to form a smooth and elegant shell; the natatorium employs cable membrane structure to create a light and transparent appearance. The four venues are unified under one green roof. To accommodate the long span, the green roof is made of two-way planar trusses, and the main truss is provided with prestressed cables to control the vertical deformation. The reinforced truss floor slabs are adopted. Through the close cooperation between architects and structural designers, a holistic, harmonious and the representative project presence is established.

惠州金山湖游泳跳水馆
Jinshan Lake Swimming and Diving Natatorium, Huizhou

项目地点：惠州市，惠城区
设计时间：2006- 2008 年
建设时间：2008- 2010 年
建筑面积：24,574 平方米
建筑师团队：郭胜、陈雄、吴冠宇、张春灵、邓弼敏、谭耀发、李健棠、韩锡超、黄科、黄珏欣
主要奖项：国家优质工程银质奖
全国优秀工程勘察设计行业奖 · 建筑工程二等奖
中国建筑学会建筑创作佳作奖

Location: Huicheng District, Huizhou City
Design: 2006-2008
Construction: 2008-2010
GFA: 24,574m²
Design Team: Guo Sheng, Chen Xiong, Wu guanyu, Zhang Chunling, Deng Bimin, Tan Yaofa, Li Jiantang, Han Xichao, Huang Ke, Huang Juexin
Major Award(s): National Quality Engineering Sliver Award, The Second Prize (Construction Engineering) of National Excellent Engineering Exploration and Design Industry Award, Honorable Mention of the ASC Architectural Creation Award

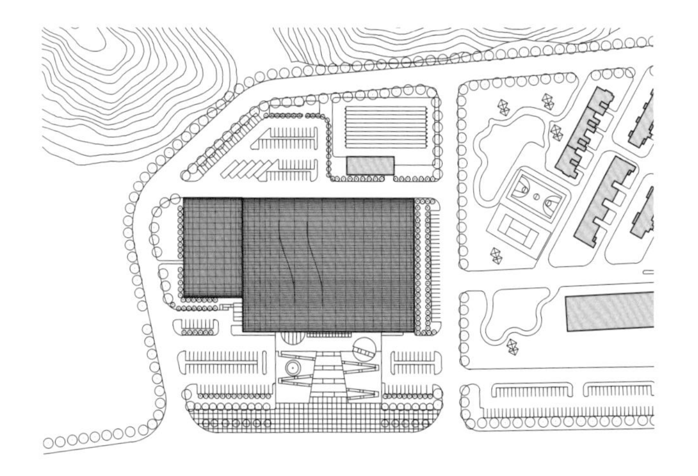

惠州金山湖游泳跳水馆是我们 2004 年成立机场所之后，于 2006 年投标中标的第一个体育建筑，是省运会的一个场馆。项目的选址也是比较可持续的，在惠州大学旁边，沿大学主要门口前马路两边，一边是体育馆，一边是游泳馆，两个项目都是我们院中标，而机场所做了游泳跳水馆。这个项目是全国的竞赛，由我们原创中标实施的。

项目在自然环境里，附近是金山湖，有山有水，自然地形成了山水题材。另外因为是游泳跳水馆，所以本身就有了很流动的水的主题，因此最后我们提出来山水意象这个主题。这是一个非线性三维的造型，是由山的主题和水的主题两个结合在一起，经过多方案的推演之后，最终形成的很具有动感的造型。项目也打破了传统体育馆表现力量感的模式，融入了更多流动和优美的造型，它既像山，与山协调，又像水，有水的灵动。

当时要实现这么复杂的造型，并没有用到犀牛，而是用 3ds Max 做的。用了切片的方式，一个个剖面切片去表达这个形体，建筑和结构配合着完成。这在全国其实也是比较早期的一个复杂的形体设计，而且完成度很高。

室内空间跟结构的配合非常到位，结构高低起伏却很干净，没有很多附加构件。采用刚架结构，没有用桁架或网架，在观众席上面两边加了“人”字形的柱支撑，再在刚架之间使用了纵向的钢构件，这样设计有效地减少了跨度，还解决了拉杆平衡纵向刚度的问题，是建筑和结构深度配合的设计。

还有一个很特别的地方，就是跳台的设计，我们做了一些椭圆的背景板，包括跳台和记分牌，与流动的室内空间很协调。跳台是经过精心设计的，一般体育建筑的跳台，从楼梯走上去，观众看得到运动员，运动员也看得到观众，但这个项目的跳台设计，运动员走上去的时候是看不到观众的，观众也看不到运动员。因此运动员一出来的时候，是一个高调的亮相，形成开启人生舞台的意境。跳台本身是用三维的清水混凝土做的，很有特色。

上二层主要的交通线有个景观台阶，旁边是“Z”字形残疾人坡道，残疾人坡道融合在景观平台上，是一个很巧妙的无障碍设计，成了整个景观的一部分。结合景观台阶设置了一些圆洞，使大台阶下面的空间能够有更好的体验感，因为这下面主要是其他的流线，包括体育运动员的流线等。

Planned for the Provincial Sports Games, the project represents the first sports building bid we won since the establishment of the airport office in 2004. The project is sustainably sited beside Huizhou University. Actually GDAD won bids for both buildings beside the road in front of the university's main entrance, namely the gymnasium on one side and the natatorium on the other. The natatorium is designed by the Airport Office which won a national design competition with original creative design proposal.

The Project is embraced by natural landscape, in particular the mountains and Jinshan Lake nearby. Such a project context naturally defines the project theme of mountains and waters. Also the natatorium itself involves the water element. So we finally proposed the project theme of mountain and water imagery . The building presents a nonlinear dynamic3D shape which combines the themes of mountains and waters through evolution of multiple options. Different from conventional gymnasium which emphasizes strength, the project brings a more flowing and graceful form which resembles and harmonizes with mountains and waters.

3ds Max instead of Rhinoceros was used to achieve such a complex form. The architects and structural designers worked closely to represent the building form via sections slicing, and successfully accomplished the task. This is an early example of complex form design in China . Interior spaces well fit the structure which fluctuates yet appears neat without superfluous components. The project employs rigid frame instead of truss or grid, adds herringbone column support to both sides of the auditorium, and provides longitudinal steel members between the rigid frames. This approach effectively reduces the span and balances the longitudinal stiffness of tie rods, reflecting in-depth collaboration between architecture and structure,

Another highlight is the platform where some elliptical backdrops are incorporated, including platform and scoreboard, to harmonize with the flowing indoor spaces. The platform is elaborately designed so that the athlete walking upstairs and the spectators cannot see each other, while the case is the opposite for a conventional platform. So the athlete is presented in a high profile on the platform, symbolizing the athlete's debut on a stage of life. The platform is distinctively featured with 3D fair-faced concrete construction.

The main circulation to F2 is planned with some landscaped steps, beside which is a Z-shaped ramp for the disabled. The ramp is integrated into the landscape platform as an ingenious barrier-free design and part of the whole landscape. Some round holes are cut along with the landscaped steps to improve the user experience in spaces below the generous steps, as these spaces accommodate other circulations like those for athletes.

会展建筑

EXPO BUILDINGS

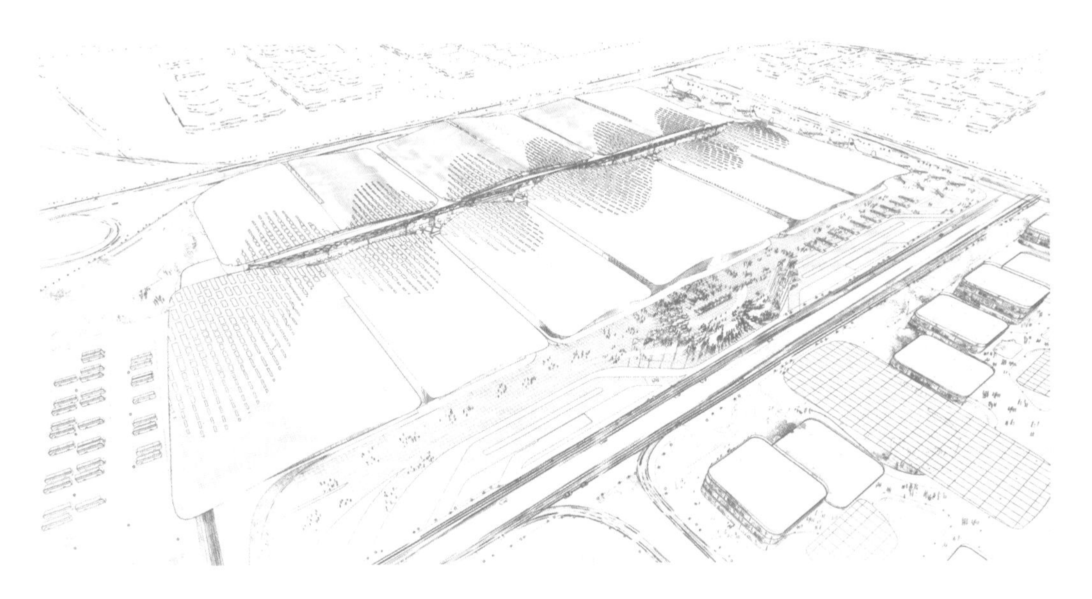

会展建筑体现出地域及人文的共有价值，为大家提供自由、平等的共享场所，力求使其成为城市活力的重要源泉。

Expo buildings highlight shared regional and cultural values, provide shared space that is free and equal, and serves as an important source of urban vitality.

广州空港商务区会展中心

Guangzhou Aerotropolis Development District (GADD) CBD Exhibition Center, Guangzhou

项目地点：广州市，白云区
设计时间：2020 年至今
建设时间：2021 年至今
建筑面积：350,000 平方米
合作单位：AXS SATOW，广州市公用事业规划设计院有限责任公司
建筑师团队（广东省建筑设计研究院有限公司）：
陈雄、黄俊华、罗若铭、高原、苏青云、龚锦鸿、卢宇、陈冠东、钟仕斌、陈俊明、陈颖、魏江峰、黄旭峰、李锦中、朱伟鑫、梁红缘、胡冰清、张熠、黄晋奕

Location: Baiyun District, Guangzhou City
Design: 2020 to present
Construction: 2021 to present
GFA: 350,000m^2
Partners: AXS SATOW INC., Guangzhou Public Utilities Planning and Design Institute Co., Ltd
GDAD Design Team: Chen Xiong, Huang Junhua, Luo Ruoming, Gao Yuan, Su Qingyun, Gong Jinhong, Lu Yu, Chen Guandong, Zhong Shibin, Chen Junming, Chen Ying, Wei Jiangfeng, Huang Xufeng, Li Jinzhong, Zhu Weixin, Liang Hongyuan, Hu Bingqing, Zhang Yi, Huang Jinyi

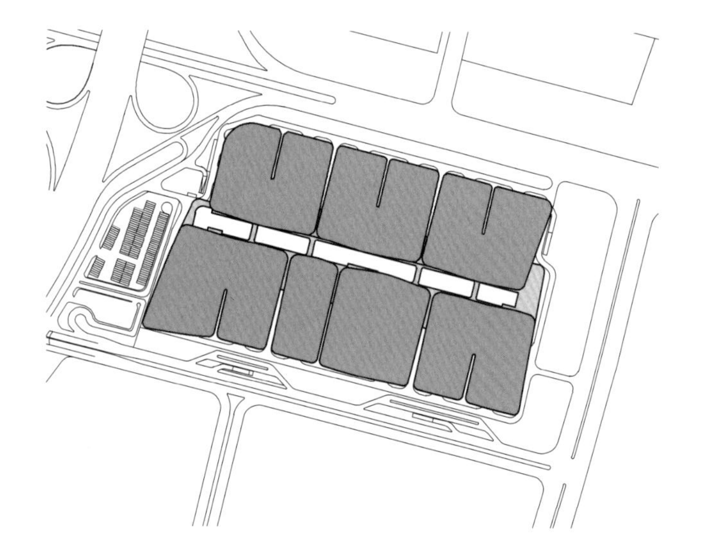

广州空港商务区会展中心位于白云机场旁边比较重要的地段。空港会展的选址与布局方式非常典型，比如国内上海虹桥枢纽与国家会展中心，联系也方便。深圳机场和深圳会展中心也是联系紧密的关系。基于白云机场的支撑，这样一个超大型的空港会展是非常好的选项。

空港会展如何体现出它的特色？一方面是会展本身的标志性，另外就是会展中心的整个布局，它是鱼骨式的场馆组织，在会展中是比较典型、有效的布局方式。

如何把用地的特征跟空港的特征结合在一起，使得整个形态能够做出特色？我们采用了两维波浪起伏的屋顶去实现。布局前后南北的场馆错开，形态与场馆的布局都比较灵动。

通过南边中间的大登陆厅，以及东西两边的小登陆厅，构成了会展这种建筑里面典型的登陆厅功能。主要的交通轴是一个“T”字形的交通轴，东西通廊以及南边主要的门厅串联在一起，基本布局就这

样确定下来。通过高低起伏的形体组合，体现流动的感觉，跟“云”的主题其实是相契合的，也表达出白云腾飞的感觉。

项目位于机场高速旁，进出白云机场的门户位置上，因此我们对于城市呼应方面作了充分考虑，以灵动的形态出现在机场路的旁边。

会展中心内部空间也是非常有特色的，东西向的主轴空间，实际上是一个半室外的通廊，有自然采光、自然通风，能够遮风挡雨。该主轴空间建构由建筑专业与结构专业进行了很充分的配合，实现了一个树状自然生长的形态。

它的整个室内包括登陆厅的设计，也通过两维的方法去做出三维灵动的空间界面，使得外壳的灵动造型跟室内空间的流动感结合在一起。每个展馆的结构跟建筑的配合也是很有特色的，屋盖结构采用梭形复合张弦桁架，整个跨度比较大，也有效地减少结构的高度，这个也是很有特色的。

展馆不分上下层，通过首层紧凑布置满足进出流线。外围以及两馆中间设有环形可回旋的货运通道，在中轴主要为步行区域，实现人货分流的交通组织。

由于项目位于空港区域，第五立面比较重要。设计结合了太阳能光伏板，按照标准板排出肌理，它是一个有表情的、非均质的光伏板布置，组成了灵动的图案，利用光伏板元素构成了丰富的第五立面。进出港的时候，无论白天或晚上都能看到这个展馆，所以灯光也结合这个图案也做了一些设计。

项目整体造型轻盈灵动，体现了空港的特色，也营造出一种开放的姿态，融入城市生产生活，成为空港门户的新地标。

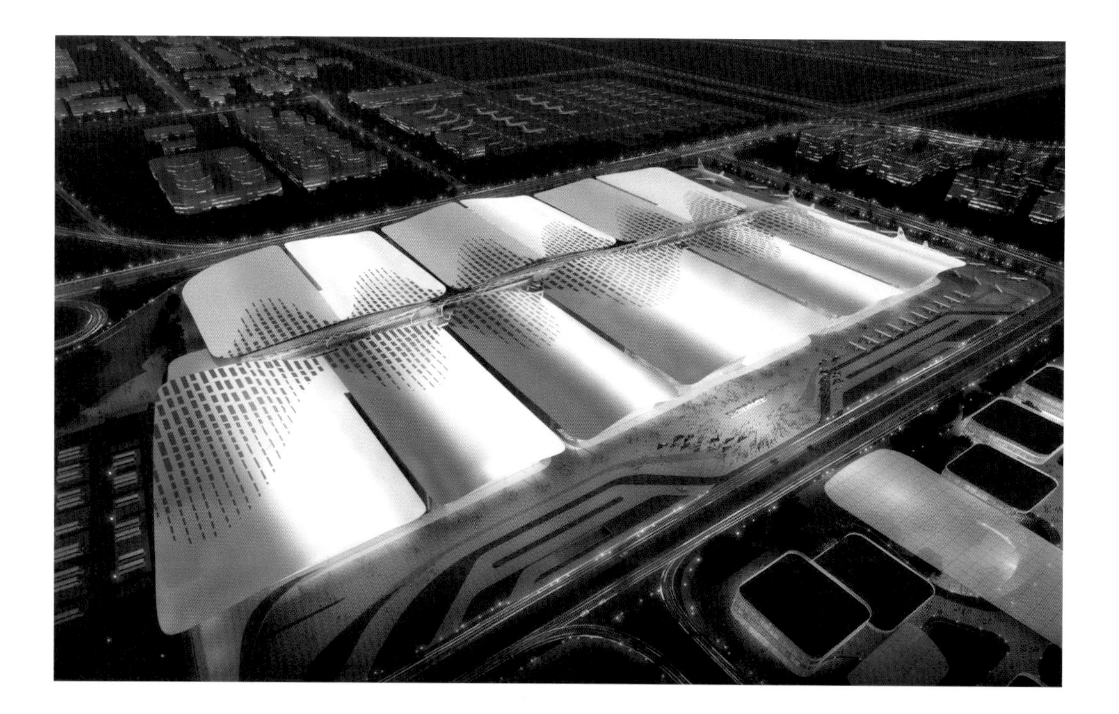

The project is prominently situated in close proximity to Baiyun Airport, which reflects a typical siting and layout of airport-based exhibition facilities, such as the conveniently connected Shanghai's Hongqiao Transportation Hub and the National Exhibition and Convention Center (NECC) Shanghai, as well as the closely related Shenzhen Airport and Shenzhen World Exhibition and Convention Center. Backed by Baiyun Airport, the project can surely demonstrate its strength as a super-large aviation exhibition center.

To reflect the characteristics of the project, we focused on the landmark identity and overall layout. The herringbone layout used is typical and effective for exhibition facilities.

To integrate the site with the airport context and create a distinctive form, we adopted a two-dimensional undulating roof design. Venues on the north and south are staggered in layout, realizing very flexible and lively building forms and layout.

The large lobby in the centre of the south, plus two smaller ones on the east and west, offer typical lobby functions for the exhibition buildings. A T-shape main traffic axis connects the east-west corridor and the main lobbies in the south, defining the basic layout of the project. The undulating building blocks are combined to create a flowing form, echoing the cloud theme and symbolizing the city's ambition.

In view of the project's location beside the Airport Expressway, a gateway to Baiyun Airport, the building form is designed to respond to the urban development, i.e. to rise lively beside the expressway.

The interior spaces are very unique. The east-west main axis space, which is actually a semi-outdoor corridor, enjoys daylight and natural ventilation, and is sheltered from wind and rain. Thanks to the close coordination between the architects and structural engineers, the spatial construction of the main axis realizes a natural form resembling the growth of trees.

In interior design, including that of the lobbies, an flexible 3D interface is created via 2D approaches, which well combines the lively building form with flowing interior spaces. In each exhibition hall, the structure is uniquely combined with the architecture. The roofing features a distinctive shuttle-like composite tensioned truss in large span which effectively reduces structural height.

The exhibition hall features a very compact layout with all incoming and outgoing circulation on a single floor. Circular freight passages are provided outside and between exhibition halls, while pedestrian traffic is planned along the central axis, realizing separated pedestrian and vehicular circulation.

The fifth facade of the project is important due to its location in GADD. Typical solar PV panels are incorporated to form the desired texture. The expressive and non-homogeneous solar panel arrangement creates lively patterns, presenting a diversified fifth facade featuring solar energy. The center is visible for arriving and departing passengers in the day or at night, so the lighting design also references the pattern of the facade.

The lively and flexible form of the project highlights the airport features, and presents an open gesture. It makes the project part of the city's development and life, and creates a new landmark at the gateway of GADD.

广东（潭洲）国际会展中心
Guangdong (Tanzhou) International Convention and Exhibition Center, Foshan

项目地点：佛山市，顺德区
设计时间：2016 年
建设时间：2016 年
建筑面积：115,549 平方米
合作单位：同济大学建筑设计研究院（集团）有限公司，广东南海国际建筑设计有限公司
建筑师团队（广东省建筑设计研究院有限公司）：
陈雄、罗志伟、钟伟华、陈艺然、梁石开、何业宏、黄亦彬、敖立、陈伟根
主要奖项：行业优秀勘察设计奖 · 优秀（公共）建筑设计二等奖
中国会展业金海豚大奖 · 中国会展标志性场馆
广东省优秀工程勘察设计奖 · 建筑工程二等奖

Location: Shunde District, Foshan City
Design: 2016
Construction: 2016
GFA: 115,549m^2
Partners: TJAD, NHCI
GDAD Design Team: Chen Xiong, Luo Zhiwei, Zhong Weihua, Chen Yiran, Liang Shikai, He Yehong, Huang Yibin, Ao Li, Chen Weigen
Major Award(s): The Second Prize of Excellent (Public) Architecture Design of National Excellent Exploration and Design Industry Award, China Exhibition Industry Golden Dolphin Awards -Landmark Venue of the Year, The Second Prize (Construction Engineering) of Excellent Engineering Exploration and Design Award of Guangdong Province

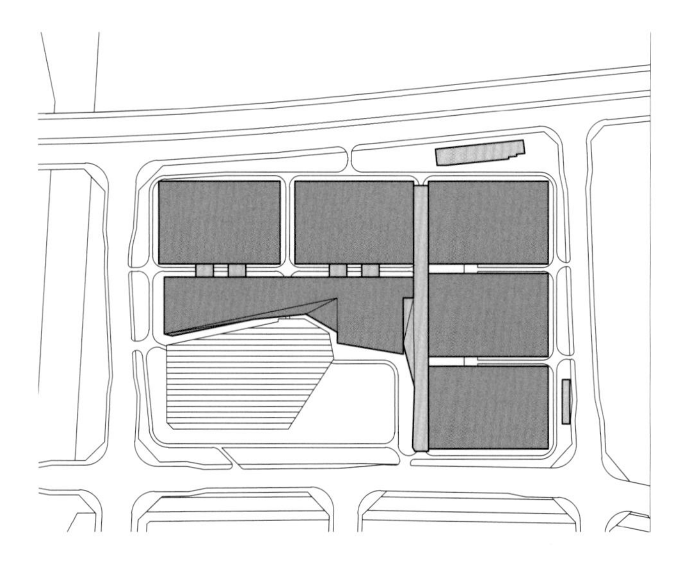

项目由五个标准展厅及一个登陆厅组成，由于整体工期比较短，因此先建五个标准展厅，后建登陆厅，满足开展的要求。五个展厅平面基本上都是标准构成，整个造型也比较标准，在这个情况下做了一些形体错动，满足快速施工的要求。项目最大的挑战是从确定方案之后的 9 个月，从设计到建造完成，是非常快速的设计和建造的过程。9 个月，从图纸到装修做完，非常具有挑战性，建筑、结构、机电、幕墙等各专业高效配合，同时跟施工推进有机地结合在一起，满足施工快速推进的要求。设计团队作了非常大的努力，最终也实现了这个目标，业主也是非常满意的。它的完成度也是比较好的，后面再进一步就是把登陆厅做好。实际上五个展厅在 9 个月之后，各方面条件满足后已经开始使用，之后做完登陆厅，功能进一步完善。

室内部分是一个大跨度的结构，我们团队作了很好的资源整合，做大跨度建筑也是我们的强项，它是一榀一榀的三角形标准桁架。天窗的部分也跟结构结合在一起，加上室内的吊顶，形成整体性很强的设计，结构与装修和建筑都结合在一起，是比较清晰的结构概念。

幕墙设有遮阳构件，作为表皮的一种做法。登陆厅是由一些折面组成，折面之间形成组合的灰空间，带来很好的体验。

潭洲会展中心项目整体完成度是比较高的，为区域产业助力，推动了佛山优势产业的发展。

The project is composed of five typical exhibition halls and one lobby. Subject to short construction period, the five halls will be built before the lobby, which will be built later, to meet the conditions for operation. The five halls, which basically follow a typical floor plan, take on a relatively regular overall form. In this circumstance, some staggered forms are made for quick construction. The biggest challenge was the 9-month construction after the design proposal was finalized, which was a very quick process from project design to completion. From drawing design to decoration completion, it took only 9 months, which was very challenging. Thanks to highly efficient cooperation and full coordination among architecture, structure, MEP, curtain wall and other disciplines during construction, the rapid construction progress was guaranteed. The design team made great efforts and eventually accomplished the goal, much to the client's satisfaction, and then they continued to complete the lobby. In fact, 9 months later with everything ready, the five exhibition halls were put into operation, and the subsequently completed lobby further complemented the functions.

The interior building space is supported by a large-span structure, which is realized through excellent resource allocation, and of course, based on our strength in making such large-span buildings. With the structure made of typical triangular trusses, the inbuilt skylight, and the indoor ceiling, we have made a very holistic design. This integration of structure with decoration and architecture creates a relatively clear structural concept.

Curtain walls are equipped with shading which serves as the building skin. The lobby features some folds with grey space in between, bringing a very pleasant experience to users.

Highly realized as per our design, the project is expected to support the local industries and drive the development of competitive ones in Foshan.

第十三届中国航展扩建工程
The 13th Airshow China Expansion Project, Zhuhai

项目地点：珠海市，金湾区
设计时间：2019- 2020 年
建设时间：2019- 2021 年
建筑面积：40,372 平方米
建筑师团队：陈雄、郭其轶、黄俊华、钟伟华 、许尧强、高原、邓章豪、金少雄

Location: Jinwan District, Zhuhai City
Design: 2019-2020
Construction: 2019-2021
GFA: 40,372m²
Design Team: Chen Xiong, Guo Qiyi, Huang Junhua, Zhong Weihua, Xu Yaoqiang, Gao Yuan, Deng Zhanghao, Jin Shaoxiong

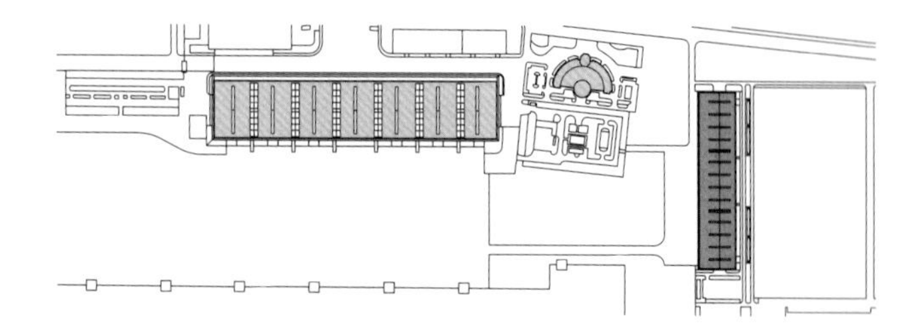

这个是 2020 年珠海航展的扩建工程，里面是有三个馆，还有室外展场一体的设计。航展有快速建造的要求，因为是三年一届，建设过程要在两届之间完成，所以从设计到施工，都有快速推进的节奏来满足它的要求。

这个项目也是大跨度的钢结构，是我们团队能够胜任的类型。它其实有点像会展的功能，只是围绕它的主题是航空产业，航空的设施设备，还有飞机的展示。所以整个项目的难点除了在于快速的推进，还在于大跨度的结构。

设计方案中，每个馆设置了一个 72 米宽、14 米高的超大尺寸门，是能够进入飞机的，像飞机库一样，有一些气控门在里面，这是团队第一次做带有气控门的工业建筑，它的标准化建造，是快速推进的基础。结构设计上也有较多创新，包括管桁架混合锥网架的展厅结构、超大桁架结构、超大地坪荷载、超大荷载管沟盖板等，满足飞机展览的要求。

这个项目满足了业主快速推进的要求，同时实现了这些功能，所以最终业主是相当满意的。

This project, which consists of three pavilions and outdoor exhibition venues, is an expansion of the 2020 Zhuhai Airshow. Since the Air Show is held every three years, the design and construction have been completed within the time interval.

This project also features a large-span steel structure, which is where our design strength lies. Sort of like convention buildings, its functional design centers on the aviation industry, including aviation facilities, and aircraft display. Therefore, the major challenge of the project lies in the long-span structure, apart from quickening the progress.

Each exhibition hall is designed with a giant air-control gate 72 m wide and 14 m high, and standardized construction of the gates provides a basis for quickened progress. There are also many innovations in structural design, including the exhibition hall structure featuring pipe truss and cone net integration, the super large truss structure, the super heavy-duty floor, the super heavy-duty pipe ditch cover, etc., to meet the requirements for aircraft exhibition.

Much to the client's satisfaction, the project successfully progressed and fulfilled its functional purposes.

开发类建筑

DEVELOPMENTS BUILDINGS

开发类建筑是城市中最大量的建筑类型，需迎合各式需求与资源禀赋，有效提升属地的综合价值。

Development buildings represent the building typology of largest quantities in cities. They need to cater to varied needs, assets and resources to effectively enhance the comprehensive value of the site.

珠海横琴保利中心
The Poly Center in Hengqin, Zhuhai

项目地点：珠海市，香洲区
设计时间：2011-2015 年
建设时间：2015-2018 年
建筑面积：220,210 平方米
合作单位：AXS SATOW
建筑师团队（广东省建筑设计研究院有限公司）：
陈雄、郭胜、陈超敏、林建康、宋定侃、黎昌荣、莫颖媚
主要奖项：行业优秀勘察设计优秀奖 ·（公共）建筑设计一等奖
中国建筑学会建筑设计奖 · 绿色生态技术一等奖
国家城市设计试点珠海优秀项目
中国威海国际建筑设计大奖赛银奖
香港建筑师学会两岸四地建筑设计大奖 · 商业类银奖（金奖空缺）

Location: Xiangzhou District, Zhuhai City
Design: 2011-2015
Construction: 2015-2018
GFA: 220,210m²
Partners: AXS SATOW INC.
GDAD Design Team: Chen Xiong, Guo Sheng, Chen Chaomin, Lin Jiankang, Song Dingkan, Li Changrong, Mo Yingmei
Major Award(s): The First Prize of Excellent (Public) Architecture Design of National Excellent Exploration and Design Industry Award, The First Prize (Green Ecological Technology) of ASC Architectural Design Award, Excellent Project Award of National Urban Design Pilot City in Zhuhai, Sliver Award of the Weihai International Architectural Design Grand Prix, Sliver Award (Business Category) (gold award left open) of Hong Kong Institute of Architects (HKIA) Cross- Strait Architectural Design Symposium and Awards

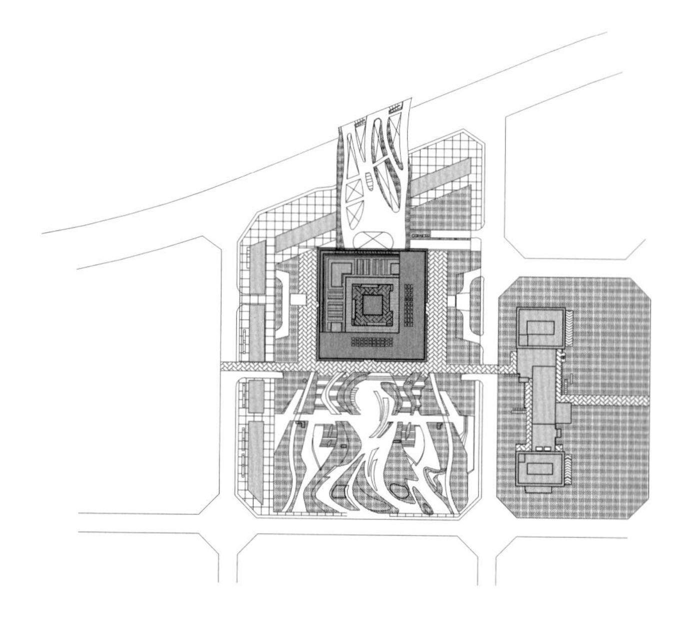

2011 年，珠海横琴这个项目进行国际招标，我们跟佐藤一起合作参与了，这是与佐藤合作投标成功的第二个项目，我们跟佐藤最早的合作是广州萝岗科技人员公寓，所以大家在一起合作，横琴又是一个新的起点。我们在联合原创方面，机场所团队展现了很好的成长过程。

机场所团队其实是多元化经营，这个是比较重要的，因为我们做超大型项目的机会相对不是那么多，加上整个项目的周期很长，因此其实从业务或者经营角度看是需要有其他类型的项目来进行补充的。从团队的成长来讲，通过一系列其他的项目，才使团队能够更全面地成长，同时这些项目快速推进，相对流程会短一些，也更有利于团队的经营和成长。当时作为所长，我除了比较关注团队的主业，同时也要关注多元化的业务类型。从另一个维度去思考这些问题，我们做大型项目的经验，是可以复制到其他项目上去的，所以在有需求的领域，多元化也是我们其中的一个方向。

最初规划新区时，项目定位为横琴区政府办公建筑。项目位于新区轴线上，背靠小横琴山，面对大横琴山，整个自然环境相当优美。中间东西朝向还保留了一些原来的水面，是在轴线上的景观。所以当时我们投标中标，想了想，这样一个形体，是政府的形象，因此述标时，

除了讲述这个方案本身的特色，还补充了三个比较重要的特点。

第一点，是它的手法。作为政府建筑，它没有传统设计的那种威严，而是亲民的。传统政府建筑大部分都有个“八”字形的坡道让汽车可以上去，加上一个大台阶，具有威严的形象。这个项目没有这些设计，也不强调轴线，更加灵动。整体形象包括入口没有中正对称的感觉，是一个很自然的形态。第二点，是它的外立面给人很有效率的感觉，因为都是线性元素，加上方正的形体，体现了秩序感，也暗示了政府的效率。第三点，它是绿色环保概念很足的项目。亲民的、有效率的、绿色环保的，这几个理念在建筑的特征都能够体现。

设计本身也是很特别的，因为我们希望能够创造横琴新区的一种风格，这种风格来自珠海作为滨海城市，蓝天白云、青山绿水的环境，因此我们用白色的线条表皮，呼应滨海城市的特质。很多滨海城市都喜欢用白色，为什么呢？因为海是蓝色的，天空也是蓝色的，用白色去衬托它是很舒服的。另外是双层表皮的应用，外表皮是非均质的遮阳百叶，内部是传统的玻璃幕墙表皮，这是很重要的特色。

它的体量很大，是 100 米 × 100 米，高度也接近 100 米的立方体。这么大型的体量，怎么创造比较好的形象和环境呢？最后是在表皮的四个方向都开了露台，而凡是在立面上可以看到的这些露台，作

为每层平面的公共区域都可以到达。表皮里还隐藏有另外一个层次的露台，立面上可以看到的露台是公共的，但是没看到的地方，也还有些露台藏在表皮下，是各自办公区的专属。有公共的，也有专属的，两层露台跟外面表皮叠合在一起，创造了非常丰富的空间体验。建筑整个形体是架空的，四个角分别设置很大的核心筒，通过巨型桁架把整个建筑托起来。桁架占两层高，上面托着，下面架空，是因为我们希望能够实现通风，城市的空间也能够穿过建筑。但是到最后把它围起来了，因为改变了功能，项目不再是政府的，通过拍地由保利开发，作为开发项目来建设，因此功能作了很多的调整，但是维持了基本的概念，白色的设计还在，架空结构的托换还在，只是下面的功能改变了，例如原来政府基座的功能都是一些展厅，还有档案馆和大小会议室，最后这些功能全部改变了，整个基座变成一个商场。

建筑的前面是南广场，它也由原来的行政广场，变成了一个很动态的自由景观组合，是立体空间的广场景观，建筑北边通过很多条灵动的高架，把平台与北边的小横琴山连在了一起。这维持了我们原来从广场一直可以走到小横琴山的基本格局，而且更加灵动。

另外一点，这个项目本身是个立方体，中间有一个大概 30 米 ×30 米的露空中庭把通风问题解决了。这个中庭没有上盖，各方向的风都能进去，是一个拔风中庭，也是我们岭南地区的通风空间的构成。这个项目体现了岭南建筑特色的几个方面，一个是对气候的有机回应，包括通风、遮阳；另外，它体现了岭南建筑与自然景观的结合，例如每层楼都有公共的露台，也有专属的露台，而且都有绿化和景观结合在一起，形成立体的开放空间。结合其他多项绿色建筑节能措施，这个项目最终取得了“三星级绿色建筑设计标识证书”。

In 2011 we teamed up with AXS SATOW INC. for the international tender for the Poly Center in Hengqin, Zhuhai, and won the bid. This was the second project we successfully bid for together, a new starting point. In terms of co-creation, the airport team/GDAD has demonstrated impressive growth.

In the planning for Hengqin New Area, the project was positioned as a government office building in the area. Located on the axis of the new area, the project was backed by Xiaohengqin Mountain and facing Dahengqin Mountain fits in the natural environment perfectly, with some of the original water surfaces retained on the axis. During the bid announcement, we added information on three more important features.

First, the caring design approach. Emphasizing neither the traditional design of large steps and splayed ramps, nor symmetrical overall layout, the project is in a natural form. Second, the linear elements on the facade. These elements and the foursquare body represent order, as well as high efficiency of the government. Last but not least, the concept of environmental protection.

In addition, we hope to create a style of Hengqin New Area, which originates from the blue sky, white clouds and green mountains and waters in Zhuhai. A double-skin system with white lines is used to represent the characteristics of the coastal city. As a critical feature, the double-skin system consists of an external skin made of heterogeneous sun-shading louvers and an internal skin made of traditional glass curtain walls. In order to ensure the consistent image of the large-volume building, patios are also designed on the four sides of the skin, which are visible on the facade. There are also some patios under the skin, which are exclusively accessible to corresponding office areas. These two types of patios are superimposed with the skin, creating abundant spatial experience.

The building is built on stilts, with large cores at the four corners and a two-story giant truss supporting the entire building for good ventilation and urban spatial permeability. But in the end, the project was handed over to Poly by the government for development, and many adjustments were made to its functions. The original halls for exhibition, archives and meetings at the base are converted into a shopping mall, and the open-up ground floor is enclosed; but the basic design concept is maintained.

The administrative square on the south side of the building is turned into a three-dimensional landscaped square, and on the north side, many elevated structures are used to connect the building with Xiaohengqin Mountain. This not only maintains the original basic layout, but also adds flexibility to it. There is an open atrium of about 30m × 30m in the middle of this project to ensure ventilation. This project reflects the characteristics of Lingnan architecture: One is the organic response to the climate, including ventilation and shading; the other is connection with nature through patios on each floor, and the integration of greening and landscape. The project also adopts a number of other energy conservation measures and obtains the Certificate of Three-Star Green Building.

广州南站核心区 TOD（B 区）
TOD of Guangzhou South Railway Station Core Area (Zone B), Guangzhou

项目地点：广州市，番禺区
设计时间：2022 年至今
建筑面积：680,000 平方米
合作单位：NIKKEN，Atelier Global，Lead8，KPF，ARUP，Mvaasia，Meinhardt，SWA
建筑师团队（广东省建筑设计研究院有限公司）：
陈雄、江刚、陈应书、郭其轶、庞熙镇、钟伟华、刘德华、何静、江祖平

Location: Panyu District, Guangzhou City
Design: 2022 to present
GFA: 680,000m²
Partners: NIKKEN, Atelier Global, Lead8, KPF, ARUP, Mvaasia, Meinhardt, SWA
Design Team (GDAD): Chen Xiong, Jiang Gang, Chen Yingshu, Guo Qiyi, Pang Xizhen, Zhong Weihua, Liu Dehua, He Jing, Jiang Zuping

这是一个很典型的 TOD 项目，它并不是跟广州南站同时设计的 TOD，而是南站使用多年以后，围绕它的使用重新完善整个片区，以及解决南站一些交通方式的转换衔接问题。这个过程的第一点，是重新梳理了交通的衔接方式，包括旅客从南站出来，便捷地完成大巴、的士、私家车、网约车等这些交通方式的换乘，这秉承了我们原来做白云机场的经验，是相通的，都是让旅客流程更加顺畅。这是一个中外合作的项目，在整个过程中，我们跟外方有很好的沟通。机场所是以大型交通建筑为主要设计类型的团队，与这个项目有非常好的契合，可以发挥出我们的经验和理念。

这么多年来高铁在快速发展，交通是解决了，但是体验这方面是一直需要提升的，所以高铁也提出会向机场的品质去发展。这一点，我觉得正好这个项目能够整合我们的经验。另外，这个项目本身是国内第一个这么贴近高铁站的 TOD 开发，所以它在东西两边跟典型的高架式高铁站的衔接是非常便捷的，一方面解决了交通问题，另一方面是在解决交通问题的基础上进行了商业、办公、酒店等功能的配套，使得整个片区的功能比较完善，充分发挥土地的价值，这样的模式对于高速发展的高铁，是一种新的 TOD 开发标杆。

这个项目跟高铁站的连接是立体的，包括地下和二层架空的部分，都是立体地与高铁站衔接。除了与高架站台比较好的衔接模式，同时也保留了原来高铁站的一些周边，包括原来两个主要进站方向的交通干道的畅通，从而使它拥有更为完善的交通体系。旅客的体验未来将会大为提升，不再像以前那样出来以后要寻找其他交通工具。

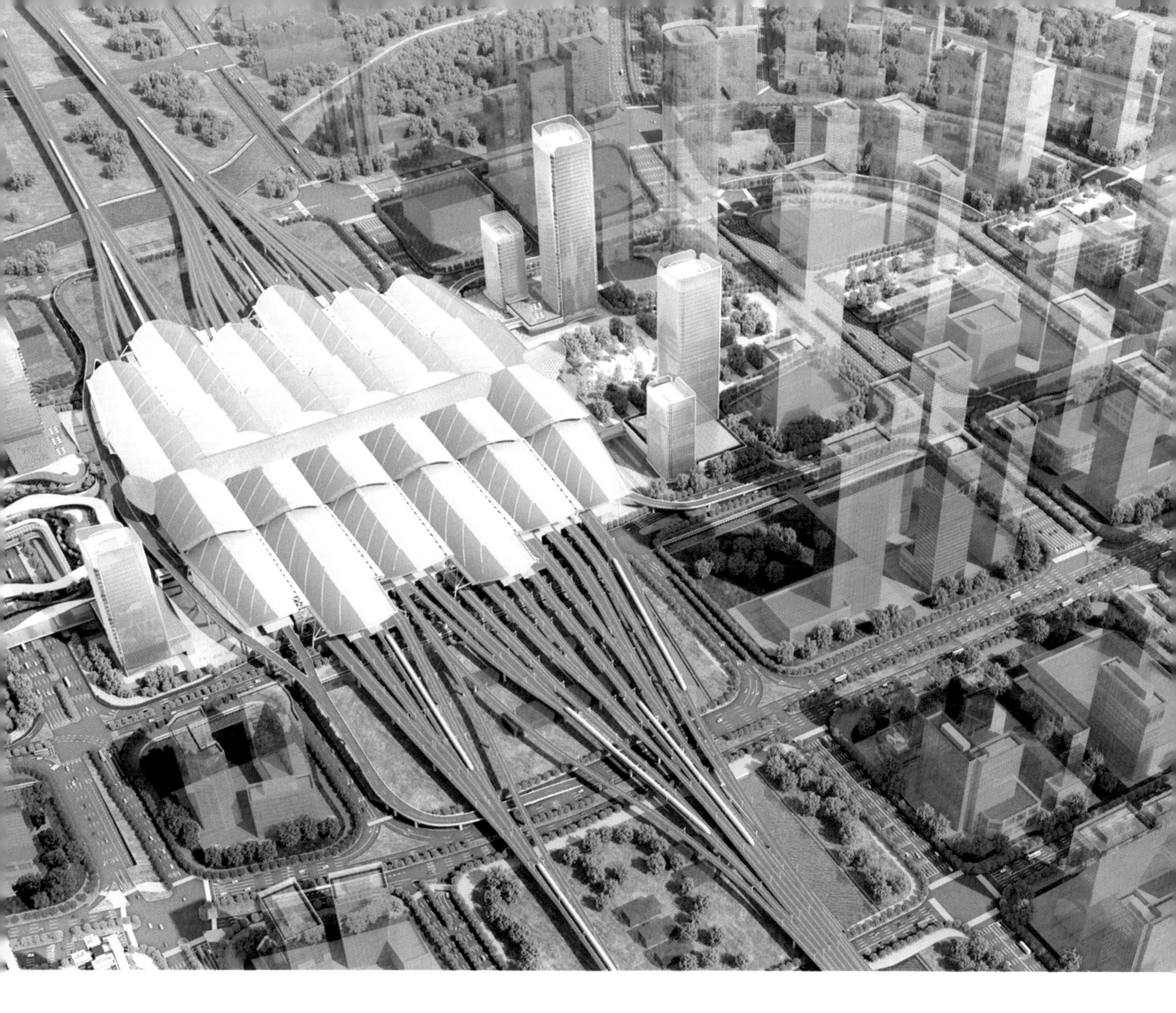

由于空间尺度比较大，对旅客来讲是比较迷茫的，所以我们TOD的开发是很明确的，一边是大巴，另一边是的士，再往前走就是私家车，基本模式其实跟机场的交通中心的模式是非常相似的。

除了交通功能的完善，还有城市客厅的打造。南站是广州一个重要的城市门户，同时由于我们的TOD开发，使它增加了城市客厅的功能，在门户中创造了休闲、消费、娱乐、商业等城市客厅的功能。这个设计不但高铁到达方便，还衔接了几条地铁线以及城际，因此所有的轨道来这里都非常方便，大流量的人流经过高铁站，会到达这些城市客厅。在这里我们希望打造一个地标式的城市客厅，不仅是对于旅客，对于整个城市的市民也能提供城市客厅的功能，这就需要打破以往我们认为高铁站只有交通为主的功能。创造这个城市客厅，除了有非常便捷的交通衔接，为了使市民愿意待在这个客厅，还需要有组织交往的功能。对于广州南站周边片区的提升，也会有非常明显的效果，因为周边片区有这个既有高铁又有轨道支撑的大型交通枢纽，是很方便的交通联系。在这个片区和周边的一些片区里，因为有如此高品质的TOD开发，与城市空间进行缝合，使整个片区能够整体地提升品质，将有效地克服以前在南站周边片区交通衔接或者城市空间比较割裂的问题。

This is a very typical transit-oriented development (TOD) project aiming to improve the area where the South Railway Station stands for many years and addresses the issues with the modes of transportation. We reorganized some connections to facilitate passengers' transfer from the station to other modes of transportation such as coach, taxi, private car and online hailing car. This is based on the same principle that we adhered to when working on the Baiyun Airport project, which is to provide a more seamless experience for passengers. This is a Chinese-foreign cooperation project where we have had good communication with our foreign counterparts. GDAD Airport Branch Institute is a team mainly focusing on the design of large-scale transportation building. Therefore, we are the perfect choice for this project where we may give full play to our experience and ideas.

Over the years, improvements in passengers' experience at high-speed railway stations have always been sought with the experience at airports as the benchmark. This is China's first TOD project that is so close to a high-speed railway station. The project makes it very convenient to leave and arrive at the typical elevated high-speed railway station on the east and west sides, resolving the issues with transportation. In addition, supporting functions such as commerce, offices and hotels are also designed, which improves the area and makes full use of the land. The project set a new benchmark for TOD for high-speed railway stations that are growing quickly. The connections between this project and the high-speed railway station are three-dimensional, including those in the underground and overhead parts. In addition to the better connection with the elevated platform, the project retains some original facilities in the surrounding areas of the station, including the smooth trunk roads in the two main entry directions, to develop a more

complete transportation system. Passengers' experience will be also greatly improved, as they no longer have to look for other means of transportation when they walk out of the station as before. Because this is a large-scale project, it will be a challenge for passengers to find the right way. That's why we designed clear paths in the TOD: one path leads to the coach zone on one side, one to the taxi zone on the other side, and one forward the private car zone. The layout is very similar to that in the transportation center of an airport.

In addition to the improvements in transportation, the project also works on the development of city parlor. The South Railway Station is an important gateway to Guangzhou. The TOD project adds some functions of a city parlor to the gateway, including recreation, consumption, entertainment and commerce. This design not only makes it convenient to access the station through the high-speed trains, but also connects several metro lines and inter-city lines. In this way, the convenient transportation will bring large flows of people to the city parlor through the station. We intend to design a landmark city parlor, not only for the passengers, but also for the citizens, which means we must think outside the box instead of seeing the high-speed railway station as merely a transport hub. In addition to convenient connections, the city parlor must also be made for social interactions to attract the citizens. This will be of great benefit for the area surrounding the station, because it is a large transportation hub easily accessible through high-speed trains and other modes of railway transportation. Because of the high-quality TOD project, the station and its surrounding area are integrated into the urban space. This improves the quality of the whole place as will be no longer disconnected with other parts of the city as before.

珠海横琴国际科技创新中心
Hengqin International Science and Technology Innovation Center, Zhuhai

项目地点：珠海市，香洲区
设计时间：2016- 2021 年
建设时间：2016- 2021 年
建筑面积：365,000 平方米
合作单位：Aedas
建筑师团队（广东省建筑设计研究院有限公司）：
郭胜、陈雄、陈应书、梁石开、江祖平、刘德华、林建康、周志明、邓载鹏、邓章豪、敖立、何叶宏、陈小东、邵国伟、庞熙镇、曾维翔、梁凯麒
主要奖项：亚太房地产大奖 · 中国公共服务建筑优胜奖
广东省注册建筑师协会广东省建筑设计奖 · 建筑方案奖公建类三等奖

Location: Xiangzhou District, Zhuhai City
Design: 2016-2021
Construction: 2016-2021
GFA: 365,000m^2
Partners: Aedas
GDAD Design Team: Guo Sheng, Chen Xiong, Chen Yingshu, Liang Shikai, Jiang Zuping, Liu Dehua, Lin Jiankang, Zhou Zhiming, Deng Zaipeng, Ao Li, He Yehong, Chen Xiaodong, Shao Guowei, Pang Xizhen, Zeng Weixiang, Liang Kaiqi
Major Award(s): Asia Pacific Property Awards (Excellence Award of Public Service Building, China), The Third Prize (Architectural Design - Public Building) of the Guangdong Architectural Design Award of Guangdong Province Registered Architect Association

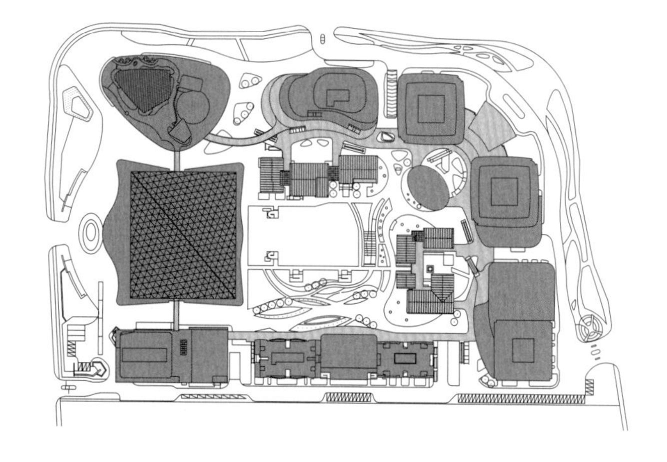

珠海横琴国际科技创新中心是我们团队与凯达合作设计的项目，是继横琴保利中心之后，我们团队在横琴的另一个比较重要的项目。这个项目也是我们机场设计团队相对多元化的其中一个方向，是可以一直持续有设计业务的产业园类型项目，这个产业园也有它的独特之处，那就是非常综合。

当时提出了一个社区的概念，因为里面有工作的地方，有住的地方，有展示的地方，还有商业的地方。以产业为主体，这是一个主题式的产业园，所以从产业的角度来看，它有比较多不同品种的产业空间，包括不同面积的产业单元，局部两层组合，因此这个项目本身是以

产业为主导的，同时创造了完善的社区功能。

项目当时是定位于中国和拉美国家合作国际论坛，因此整体设计带有拉丁美洲的建筑特色，在建筑上有饱和度比较高的色彩，而且这些标志性的符号色彩，能够结合并且赋予设计以个性。

这个项目还与岭南建筑非常有联系，适应当地气候，包括不同层面的立体花园，也包括裙楼部分的岭南风格的商业街。用了锅耳房，是岭南建筑传统民居风格的裙房和商业活动空间。这使它既能够有对外交流的印记，同时也有岭南传统文化建筑的符号。所以整个空间是具有多样性而且非常独特的，这种多样性在产业园里是比较有个性的。

我们在横琴片区有相对比较多的项目，从最早的横琴保利中心开始，国际竞赛中标之后，逐步地在这个片区承接了一批项目。国家非常重视横琴片区的发展，因为其能够与中国澳门和葡语系的国家结合在一起。有了把产业文化结合的定位，我们也顺应着横琴片区的发展去开拓业务。这个项目是我们机场所品质比较好的产业园区和办公建筑群，虽然它不是以交通为主，但它反映了我们对建筑品质的追求，这种追求获得了政府层面的认可。

Hengqin International Science and Technology Innovation Center in Zhuhai is a project on which we cooperated with Aedas Limited, another important project of our team after the Poly Center in Hengqin, Zhuhai. This project is also one of the diversified areas in which our airport design team are specialized, namely industrial park. The design of this industrial park is very unique because it involves many things.

At that time, the concept of a community was proposed, where there are places to work, to live, to show, and to do business. The project is a themed industrial park mainly accommodating industries. Therefore, it has many different types of industrial space, including industrial units of different sizes and partial two-story combined structures, supplemented by other community functions.

That was the positioning of the project determined in an international forum for cooperation between China and Latin American countries. That's why the design incorporates some architectural features of Latin America, for example, fully saturated iconic colors that make the buildings unique.

This project also incorporates the features of Lingnan architecture to adapt to the local climate, including layered gardens at different levels, as well as some Lingnan-style commercial streets among the podiums. Those are buildings with the parapets looking like the ears of a pot, podiums and places with the style of traditional Lingnan residential buildings for commercial activities. As a result, the project boasts the symbols of both foreign exchanges and traditional Lingnan architecture. Such diversity makes the industrial park a very unique existence.

Since we won the bid for the Poly Center in Hengqin, we have gradually undertaken many more projects in Hengqin area. The central government attaches great importance to the development of this area, because it ties Macao with Portuguese-speaking countries. Based on the combination of industry and culture, we have been expanding business in Hengqin area. This project boasts a high-quality industrial park and office building complex. Although it is not mainly built for transportation, it reflects our pursuit of architectural quality, which is recognized by the government.

珠海横琴科学城（二期）标段二
Hengqin Science City (Phase II) Bid Section II, Zhuhai

项目地点：珠海市，香洲区
设计时间：2019- 2022 年
建设时间：2019 年至今
建筑面积：400,000 平方米
合作单位：GMP
建筑师团队（广东省建筑设计研究院有限公司）：
郭胜、陈雄、陈应书、何静、江祖平、邓载鹏、何叶宏、罗志伟、陈小东、李岱松、廖言佳、黎昌荣

Location: Xiangzhou District, Zhuhai City
Design: 2019-2022
Construction: 2019 to present
GFA: 400,000m²
Partners: GMP
GDAD Design Team: Guo Sheng, Chen Xiong, Chen Yingshu, He Jing, Jiang Zuping, Deng Zaipeng, He Yehong, Luo Zhiwei, Chen Xiaodong, Li Daisong, Liao Yanjia, Li Changrong

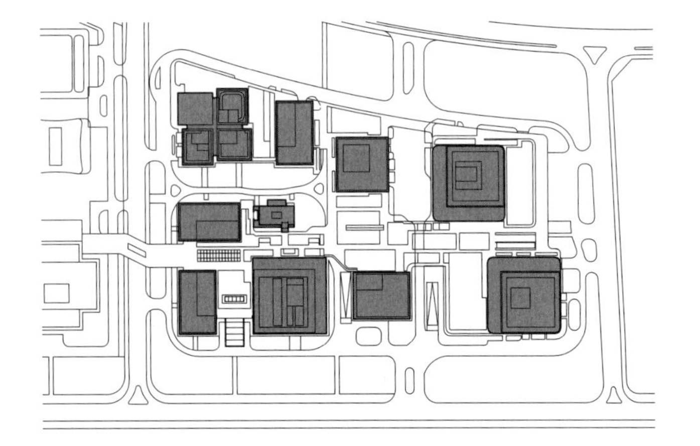

珠海横琴科学城（二期）标段二项目是团队与中建八局、GMP 的一个中外合作 EPC 项目。项目于 2020 年 3 月中标，是继横琴国际科技创新中心之后团队又一个产业园项目，也可以说是此类型项目的一个延续，也是团队在粤港澳横琴深合区的在设计业务上的一个延续。项目目前正在紧张的施工过程中，预计 2022 年年底可以主体封顶，计划 2025 年整体竣工。

项目北侧与标段一项目相望，南侧为城市绿廊，东、南和西三个方向均无高楼遮挡，向西可以远眺磨刀门水道及海景，东南向右为小横琴山，东向远望可顺马溜洲水道至十字门商务中心及澳门。基于其优越的地理位置和优惠的国家政策，横琴将推动对澳对港深度合作，成为粤港澳大湾区经济资源聚集中心和深度合作的重要平台。

强调各功能体之间的共享和联系，园区建筑通过组团方式沿基地周边布置，在南北地块的中心区分别形成各自的广场庭院空间，穿梭于建筑群的蓝绿景观带将它们连接，并通过建筑组团间的空间向场地周边渗透。

通过城市客厅的概念组团建筑，使企业交互共生。建筑组团间中央设两层通高的“城市客厅”，提供人员集散、等候休息、交流分享的公共空间，形成开放共享而相对独立的产业组团，提升园区活力，为每个建筑组团创造社区办公和生活的氛围。地面的广场庭院和空中的云平台通过几处大台阶竖向连接，台阶本身可作为公众活动的看台。

建筑组团的首层和二层为入口和商业空间，围绕中央的“城市客厅”形成建筑基座。云平台的三层为所有建筑提供了休闲活动区，同时为每栋建筑创造了复式入口空间。步行流线在地面庭院层以及三层云平台层水平方向展开，通过多处大台阶进行竖向连接，从而形成立体交织的步行系统。云平台不仅为园区提供了一个独特的绿色活动平台，同时为各建筑在首层提供了风雨无阻的畅通连接。注重自然通风、遮阳以及露台空间的设计，来自于岭南地区建筑与自然相互融合的传统智慧。

在此起彼伏的建筑群体中，以超高层建筑，分别位于场地的北端，南端和穿过场地中间，在整个建筑群体中起到引领全局的作用，建立了建筑空间的秩序感。不同的建筑高度精心组合布局，考虑各个角度的视线通道，以及从南北两个方向经过高速公路时面向城市的群体形象。

交通组织基于人车分流原则，场地机动车出入口、建筑主要入口落客港湾、机动车坡道和相应的流线都组织在建筑面向街道一侧。每个建筑组团在城市客厅处设一个落客点，机动车流线清晰，尽量降低对场地的干扰，为广场庭院和水系等休闲空间创造有利条件。场地出入口集中设置，也便于管理。

采用简洁的几何形体塑造建筑形象，虚实交错。建筑在不同高度，均有开放的空中大堂和花园，成为面向城市的窗口，也使园区内的建筑彼此形成良好的对话关系。超高层建筑大多两两成组，高低错落，形成双塔效应。每组建筑和单体在空间造型和外观设计中具有各自鲜明的特征，园区整体建筑呈现出多样统一性。

建筑内部空间和竖向划分节奏一目了然。建筑空间和形体的生成都是基于统一的格网体系，具有清晰的逻辑，实现土地和空间的高效利用。

Hengqin Science City (Phase II) Section II in Zhuhai is an EPC project on which our team cooperated with China Construction Eighth Engineering Division Corp., Ltd. and gmp. We won the bid for the project in March 2020. It is another industrial park project of our team after Hengqin International Science and Technology Innovation Center, which is also a continuation of design business in Guangdong-Macao In-Depth Cooperation Zone in Hengqin. The project is currently under construction, and it is expected that the main buildings will be completed by the end of this year and the entire project in 2025.

The project faces Section I on the north side and the urban green corridor on the south. There are no view-blocking high-rise buildings on the east, south and west, and from the project, you may see the Modaomen waterway and the sea view on the west, Xiaohengqin Mountain in the southeast, and Maliuzhou Waterway, Shizimen CBD and Macao in the east. Hengqin, enjoying a superior geographical location and favorable national policies, will promote in-depth cooperation with Macao and Hong Kong, and become a hub that gathers economic resources and an important platform for in-depth cooperation in Guangdong-Hong Kong-Macao Greater Bay Area.

In order to highlight the connection between functional groups, the buildings in the park are arranged in clusters along the site. Squares are formed in the central areas of the north and south plots respectively. The blue and green landscape belts run through and connect the building clusters to their surrounding areas.

The concept of city parlor is used in building clusters to facilitate the communication among enterprises. In the center of the building cluster, there is a two-story "city parlor" serving as a public space for gathering, waiting, resting, communicating and sharing. This will create an open, shared and independent industrial cluster suitable for working and living in the park.

The first and second floors of the building cluster are the entrance and commercial space respectively, forming the building base around the "city parlor" in the center. The cloud platform provides a recreation area for all Shizue buildings, and serves as the entrance to a layered and interwoven pedestrian system. Buildings that are different in height are carefully arranged to form sight line corridors as part of the city image.

Regarding the traffic organization, the principle of separating pedestrians and vehicles is followed. The vehicle entrance and exit of the site, the main entrances and drop-off areas of the buildings, the vehicle ramps and the corresponding circulations are all arranged on the sides of the buildings facing the street. Each building cluster has a drop-off area at the city parlor. The vehicle circulations are clear enough to minimize the disturbance to the site and create favorable conditions for recreation spaces such as squares and water systems. The centralized entrance and exits of the site are also easy to manage.

Simple geometric shapes are used to create the architectural image with solid and void elements. Buildings at different heights in the park have open sky lobbies and sky gardens that harmoniously unite the buildings. The diverse buildings in the park present unity as a whole. The interior space and the vertical division of the buildings is clear. The architectural space and shape are based on a unified grid system with clear logic for efficient use of land and space.

香港新福港地产 · 佛山新福港广场
SFK Properties · SFK Plaza, Foshan

项目地点：佛山市，禅城区
设计时间：2010- 2016 年
建设时间：2014- 2019 年
建筑面积：550,000 平方米
建筑师团队：陈雄、郭胜、梁石开、庞熙镇、程蕾、孙家明、李滔、江祖平
主要奖项：行业优秀勘察设计奖 · 优秀住宅与住宅小区设计一等奖
广东省优秀工程勘察设计奖 · 住宅与住宅小区二等奖

Location: Chancheng District, Foshan City
Design: 2010-2016
Construction: 2014-2019
GFA: 550,000m²
Design Team: Chen Xiong, Guo Sheng, Liang Shikai, Pang Xizhen, Cheng Lei, Sun Jiaming,
Li Tao, Jiang Zuping
Major Award(s): The First Prize (Excellent Residential Building and Quarter) of National Excellent Exploration and Design Industry Award, The Second Prize (Residential Building and Quarter) of Excellent Engineering Exploration and Design Award of Guangdong Province

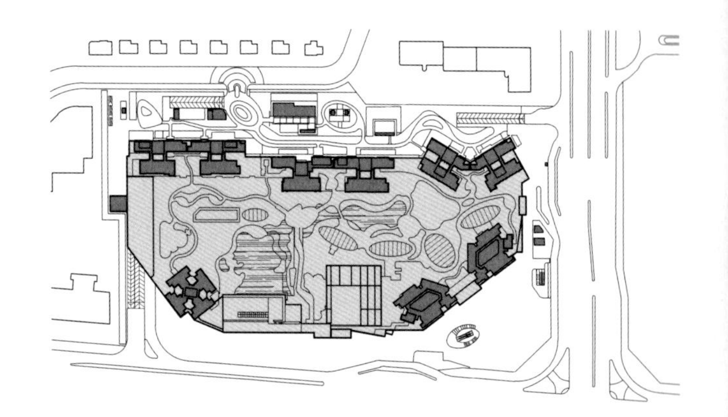

香港新福港地产 · 佛山新福港广场是一个典型的 TOD 开发项目，有两条地铁线在这里交汇换乘，与整座建筑的联系非常方便，特别是从佛山一号线进来，从地铁站出来，在负二层可以搭乘扶梯到负一层，然后从负一到地面四层都是商业，是一个非常方便的地铁上盖物业的开发。另外它的裙楼不单有商业，还有一个长途汽车客运站，其实这块地原本就是客运站，把它作为综合体进行开发，大为提升了用地的经济价值。

整个设计是比较复杂的，首先是要解决与地铁的衔接问题。建筑的交通线除了商业的组织，还要有服务长途客运站的功能，所以裙楼的出入口和交通组织等的设计都是比较复杂的。另外，在地下负二层到负四层设计车库，这样可以有效地和商业配套，并满足上面住宅的需求。

整个住宅部分位于裙楼上，在第五层做了一个很大型的屋面花园，一个人工地景花园。为了解决高密度和高强度的开发，我们非常精心地打造了一个很好的屋面花园。在这个屋面花园上共有九栋楼，主要以住宅为主，还有部分公寓和酒店，是一个高强度开发的双地铁 TOD 大型综合体。户型也是比较有特色的，因为户型类型比较多，而且当时受到国家政策调整对大户型的限制，我们采用了可以灵活拼合的设计。

整体设计还有一个比较重要的特色，就是建筑造型的风格是比较现代简洁的，趋向于公共建筑的形象，但看上去与居住功能是很贴合的，虽然没有像传统住宅的那种形态，而是相对公建化的一种立面，但是所谓公建化的立面也是很节制的，具有很干净、很洗炼的风格，这是项目很重要的一个特点。现在很多公建化的立面有太多的玻璃，但这个项目不是，它既不会有过多的玻璃，也不像传统住宅有很多的装饰线条，是比较现代简约的风格。我觉得这样的风格是可以推崇的，简洁大气，居住的品质也会比较好。还有飘窗和大阳台的设计，也是做得比较舒服的。花园上面还有一个游泳池，以及会所这些功能都集中在一起，所以从城市的角度来看，它的密度还是比较高的，基本上剩下的空地不多，主要设计了屋面上的人工地景，很花心思。

The project of SFK Properties · SFK Plaza, Foshan is a typical transit-oriented development (TOD) project, where the interchange station of two metro lines is located. This makes it very convenient to access the buildings, especially for people coming from Foshan Line 1. You may get out of the train, take the escalator from B2 to B1. From B1 to F4, there are commercial spaces easily accessible above the metro station. In addition to the commercial space, there is a coach station in the podium. On fact, the coach station has always been on the site, and is developed as a complex to greatly increase the economic value of the site.

The design is complicated, and the first issue to address is the connection with the metro station. The circulations in the building should serve both the commercial space and the coach station. That's why the design of the entrances/exits and traffic organization of the podiums is complicated. In addition, it is required to design garages on levels from B2 to B4 as facilities supporting the commercial space and the residential buildings above.

The residential buildings are above the podium. On F5, a large landscaped roof garden is carefully designed for such a high-density and intense development project. Above the roof garden, there are a total of 9 buildings, mainly residential buildings, as well as some apartments and hotels. It is a large-scale TOD complex accessible to two metro stations. There are many special house types. Because the national policy placed restrictions on large apartments at that time, we designed house types that can be flexibly combined.

Another important feature of the design is the architectural style. The buildings are modern and simple, looking like public buildings instead of traditional residential buildings, but suitable for living. Now too much glass is used on many public facades, but not this project. It does not use too much glass, nor too many decorative lines like traditional residential buildings. It adopts a modern and a simple style. I believe it is worthwhile to advocate such a simple and elegant style for good quality of residential buildings. In addition, we have also done a great job in designing bay windows and large balconies. Above the garden there is a swimming pool and a club. From the perspective of the city, the density of the project is pretty high, and there are not many open spaces left, as a result, we have carefully designed an artificial landscape on the roof.

肇庆新区喜来登酒店
Hotel Sheraton Zhaoqing
Dinghu, Zhaoqing

项目地点：肇庆市，鼎湖区
设计时间：2017 年
建设时间：2017－2021 年
建筑面积：76,000 平方米
建筑师团队：郭胜、陈雄、陈超敏、刘德华、陈伟根、邓载鹏、韦锡艳、黎昌荣

Location: Dinghu District, Zhaoqing City
Design: 2017
Construction: 2017-2021
GFA: 76,000m²
Design Team: Guo Sheng, Chen Xiong, Chen Chaomin, Liu Dehua, Chen Weigen, Deng Zaipeng, Wei Xiyan, Li Changrong

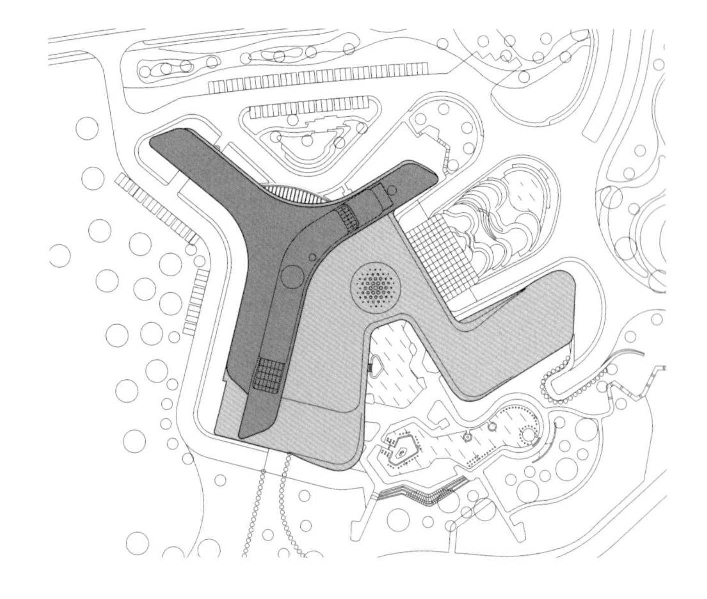

肇庆喜来登酒店项目位于肇庆新区，团队之前做了肇庆新区体育中心，后续做了肇庆东站高铁轨道交通设施，之后就完成了喜来登酒店这个区内的核心酒店项目，再后还完成了一个景观塔项目，相当于在这个新区里覆盖了好几个标志性项目。本酒店项目是比较少有的由国内本土设计团队从方案的原创开始，到施工图，再到幕墙和景观的全流程设计。项目的酒店管理方是万豪集团比较高端的喜来登品牌，建成以后成为肇庆片区档次最高的五星级酒店。

项目用地的景观条件非常好，即使在肇庆市这样一个风景城市中都是很突出的，主要的景观面向长利涌，长利涌是一条蜿蜒穿过新区的悬河，婉转悠长灵动，因此酒店采用“Y”字形的布局，一方面是最大化地争取了自然景观资源，另一方面也利用了项目地形的高差，进行了立体场地设计处理，包括主入口景观，以及长利涌周边的景观资源，都是一体化的立体开发。

酒店构型有独立的宴会厅出入口，对酒店的功能来讲是比较好的功能分区。酒店的立面造型是比较整体的，结合突出的景观阳台，形成了非常流畅而立体的水平线条。

设计团队花了很多心思营造酒店主入口的空间氛围，整个入口使用了 GRG 技术来建构树状的柱子，树状柱子上部舒展，与菱形格栅吊顶形成一体化设计，创造了非常有自然特色且大气磅礴的入口形象。

“扬帆起航，乘风破浪”，其作为万豪酒店集团中轻奢风格的酒店，采用简约时尚的设计以符合年轻一代的审美。柔美的建筑形体、通透明朗的青灰色玻璃幕墙与浅色调石材幕墙相互对比协调，融入肇庆鼎湖的青山绿水，成为一个立体景观绿色建筑。

SHERATON
SHERATON
SHERATON

Hotel Sheraton Zhaoqing Dinghu project is located in Zhaoqing New Area. This project is the third project designed by our team in Zhaoqing New Area following the Sports Center of Zhaoqing New Area and the high-speed rail transit facilities of Zhaoqing East Station. After the Sheraton, GDAD completed another sightseeing tower project. In other words, GDAD has designed several iconic projects in the New Area. The hotel project is one of the few cases where the domestic design team takes charge of the whole process of design from schematic design, to construction drawing, to the design of curtain walls and landscape. The client of the project is Sheraton, a high-end brand of the Marriott. The hotel has become the highest-rated five-star hotel in Zhaoqing since its completion.

The project site boasts superior conditions for landscape design, even in a scenic city like Zhaoqing. The main landscape faces Changli River, a perched river winding through the New Area. Therefore, a Y-shaped layout is designed to make maximum use of the natural landscape resources and utilize the height difference in landforms for the integrated development of the landscape at the main entrance and the landscape resources around Changli River.

The hotel has independent entrance and exit to the banquet hall, which is a great functional partition. The hotel facade is consistently designed with prominent landscaped balconies that create smooth horizontal lines.

The design team puts a lot of thoughts into the design of the space at the main entrance of the hotel. GRG technology is used to construct tree-shaped columns which spread in the upper part and integrated with the diamond-shaped grille ceiling to form a natural and majestic space at the entrance.

As a light luxury hotel of Marriott, the hotel will "set sail and ride the wind and waves" to attract the younger generation with a simple and fashionable design. The soft and beautiful architectural form, the transparent blue-gray glass curtain wall and the light-toned stone curtain wall are in contrast and coordinated with each other. The entire project harmoniously fits into the beautiful natural environment of Dinghu Lake as a landscaped building.

东莞中心区海德广场
Dongguan Commercial Center Zone F (Haide Plaza), Dongguan

项目地点：东莞市，南城区
设计时间：2005-2008 年
建设时间：2007-2013 年
建筑面积：217,000 平方米
建筑师团队：潘勇、陈雄、肖苑、黄科、罗志伟、李琦真、谭杨威、陈颖、李健棠、黄珏欣、黄蕴
主要奖项：全国优秀工程勘察设计奖 · 建筑工程二等奖
香港建筑师学会两岸四地建筑设计大奖卓越奖
广东省优秀工程勘察设计奖 · 工程设计一等奖

Location: Nancheng District, Dongguan City
Design: 2005-2008
Construction: 2007-2013
GFA: 217,000m^2
Design Team: Pan Yong, Chen Xiong, Xiao Yuan, Huang Ke, Luo Zhiwei, Li Qizhen, Tan Yangwei, Chen Ying, Li Jiantang, Huang Juexin, Huang Yun
Major Award(s): The Second Prize (Construction Engineering) of National Excellent Engineering Exploration and Design Award, Nominated Award of Hong Kong Institute of Architects (HKIA) Cross-Strait Architectural Design Symposium and Awards, The First Prize (Engineering Design) of Excellent Engineering Exploration and Design Award of Guangdong Province

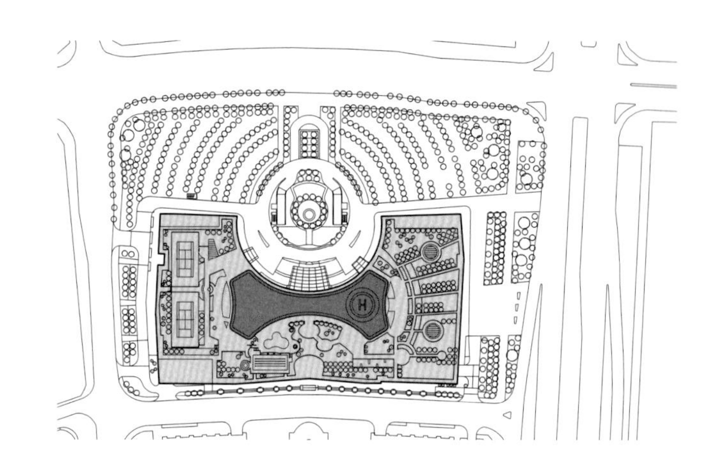

海德广场位于东莞南城区一个新规划的商业核心，在片区内包括东莞市政府、会展中心以及台商大厦。该片区是东莞市重点开发区域，项目地块恰好位于建筑群轴线端部。这条轴线长度超过 1 公里，海德广场就是该轴线端部的收官之作，因此建筑引入了“城市之门”的意向，且建筑的整体造型在当时是比较新颖的。

与海德广场相邻的会展中心造型非常的特别，整个屋面是五片梭形的叠加，给人非常锋利的感觉，带来很强的冲击感和威慑感。项目的业主考虑怎样去化解这个局面，因此联系了很多中外建筑事务所做了几十版方案，都最终都没有得到满意的结果。

我们团队介入的时间是 2005 年，当时还是机场所初创的阶段，在去现场了解场地条件和业主需求后，我们勾勒出一个初步的方案。首先是呼应建筑群轴线的“城市之门”概念，其次是用一个弧形的主楼平面，以柔克刚地化解了会展锋利屋面带来的冲击感，再有就是主楼弧形的体量和裙楼的结合，有一种太师椅般的稳重感。最终业主对我们的方案非常满意，我们的设计得以推进和实施。

海德广场的主楼是两栋“Y”字形双塔，在平面上组成了一个两端宽、中间瘦的腰鼓形。两栋主楼在顶端做了连体，形成了“城市之门”的形态。整个建筑高度约 160 米，建筑尺度结合造型和用地区位，成了整个城市片区的重要地标。

海德广场有两栋主楼，一栋主要是五星酒店，另一栋是甲级写字楼，两栋主楼顶部做了连体，通过建筑体量的挖掘，主楼之间的缝隙形成了一个“1”字的形态，呼应业主东莞第一国际商业区的命名。

项目的场地设计围绕双塔弧线形成的虚拟圆心，类似“太阳”的概念，场地内各种景观要素以及停车场等各类设施都围绕这个“太阳”的放射性而展开，形成一层一层的光环韵律。塔楼正立面面向中央广场的中心是酒店的入口，旁边则是甲级写字楼的独立入口，另外一侧的裙楼部分是宴会厅，商业在两侧和背后的裙楼，充分利用了场地资源。

Haide Plaza is located in a newly planned commercial center in Nancheng District, Dongguan, an area that also accommodates the Dongguan Municipal Government, Convention and Exhibition Center and Taiwanese Businessman Building. This is a key development area that sits at the end of the building cluster axis of over one kilometer long. As Haide Plaza is the last project at the end of the axis, the architectural design incorporates the concept of City Gateway, and the overall architectural form was a novel at that time.

The Convention and Exhibition Center adjacent to Haide Plaza has a very special shape. The entire roof consists of five superimposed fusiform structures that look sharp, impactful and deterrent. The client requested dozens of proposals from many domestic and overseas architectural firms to defuse the aggressiveness of the Convention and Exhibition Center, but none of them satisfied the client's needs.

Our airport team got involved in 2005, when the team was still in its infancy. After visiting the site and understanding the client's need, we drew up a preliminary proposal. Firstly, the design takes into consideration the concept of the City Gateway on the building cluster axis; secondly, arc-shaped planes are designed for the main buildings to resolve the impact of the sharp roof of the Convention and Exhibition Center; last but not least, the design combines the main buildings of a arched-shape volume with podiums, which are as stable as "a Grand Preceptor armchair, a traditional wooden armchair with a wide seat." The client was very satisfied with the proposal, and the design proceeded and was implemented.

The main buildings are Y-shaped twin towers. On the plane, the buildings present a waist drum shape with wide ends and thin waist. The two main buildings are connected at the top, forming the shape of a city gateway. With respect to the architectural scale, architectural form and site location, the 160-meter-high buildings have become an important landmark in the urban area. The two main buildings are a five-star hotel and a Grade A office building. They are connected at the top , while the gap in between resembles the arabic numeral 1 to echo the client's naming of the project as Dongguan First international Business District. The client's project position of the No. 1 international business district in Dongguan.

A virtual circle is designed on the site, with the arc-shape planes of the twin towers as part of the rim. The circle is like the "sun", and all landscape elements and facilities such as parking lots on the site are arranged around the radiating "sun". The center of the front facade of the tower facing the central square is the hotel entrance, and next to it is the independent entrance of the Grade A office building. The podium on the other side accommodates a banquet hall, and the podiums on both sides and the back are commercial space. All in all, the site is made full use of.

城市更新与乡村振兴

URBAN RENEWAL AND RURAL REVITALIZATION

城市更新与乡村振兴是新时代的发展趋势，通过有效的设计实践，实现功能提升与品质改善，进行文化传承与发展，创造适合人民美好生活的场所。

Urban renewal and rural revitalization are the development trends of the new era. Through effective design practice, they will enhance the building functions and quality, promote cultural inheritance and development, and create places for people's better life.

南粤古驿道梅岭驿站

Meiling Station of South China Historical Trail, Shaoguan

项目地点：韶关市，南雄市
设计时间：2018- 2019 年
建设时间：2019- 2020 年
建筑面积：514 平方米
建筑师团队：陈雄、黄俊华、郭其轶、许尧强、李珊珊、龚锦鸿、金少雄
主要奖项：行业优秀勘察设计奖 · 优秀（公共）建筑设计二等奖
中国威海国际建筑设计大奖优秀奖
广东省注册建筑师协会广东省建筑设计奖 · 建筑方案奖公建类一等奖
广东省优秀工程勘察设计奖 · 传统（岭南）建筑一等奖
广东省“三师”专业志愿者服务（勘察设计）优秀项目推荐

Location: Nanxiong District, Shaoguan City
Design: 2018-2019
Construction: 2019-2020
GFA: 514m²
Design Team: Chen Xiong, Huang Junhua, Guo Qiyi, Xu Yaoqiang, Li Shanshan, Gong Jinhong, Jin Shaoxiong
Major Award(s): The Second Prize of Excellent (Public) Architecture Design of National Excellent Exploration and Design Industry Award, Excellence Award of the Weihai International Architectural Design Grand Prix, The First Prize (Architectural Design - Public Building) of the Guangdong Architectural Design Award of Guangdong Province Registered Architect Association, The First Prize (Traditional Lingnan Architecture Design) of Excellent Engineering Exploration and Design Award of Guangdong Province, Recommended Excellent Project (Exploration and Design) of Planner, Architect and Engineer Volunteer Association of Guangdong Province

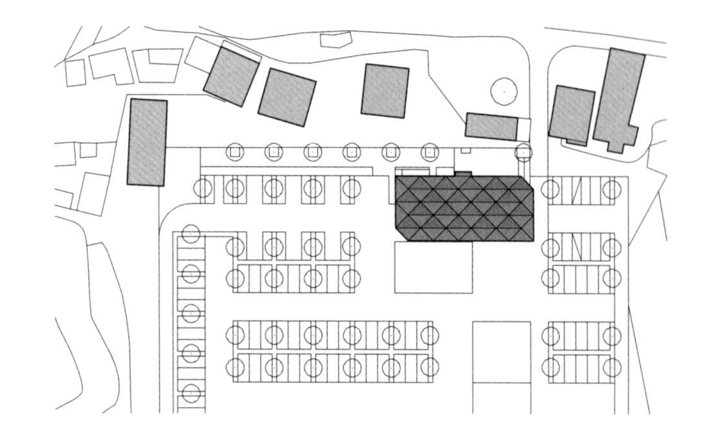

梅岭驿站是一个只有几百平方米的小项目，位于南粤古驿道这个广东历史悠久的线性文化遗产旁边，这个地方在现代来讲是经济欠发达的地区。我们当时将驿站选在梅关古道，这是广东非常出名的旅游区。

场地选在了景区的一个大型停车场旁边，靠近梅岭古村的村口。当时在选址时也有故事，我们在现有场地附近还选了几个点，原来的选址可能要拆除一些村民的房子，村民不愿意，觉得将来无法再建高楼了，所以我们最后选择了现在这个场地，与村民的小商铺、小餐饮靠在一起，可以减少扰民。这个想法得到当地政府和省政府领导的肯定，当地村民也很高兴。

项目本身的功能很简单，只有两层，我们希望建筑具有多种功能，既能作为停车场的附属设施，为旅客提供服务，又能作为村民日常休息和交往活动的场所。首层作为景区售票、宣传的地方，二层做了一些半室外休息空间，可以看山景，也可以做宣传活动，具有功能的不确定性，比较确定的是我们希望创造一个灰空间，符合岭南地区的气候，能给游客、村民提供一个多样活动的场所。

为了呈现地方的特色，我们到当地进行采风，探寻了当地建筑会出现的构造与细节，最后在驿道的北面设置了挑出阳台，用钢结构进行支撑，体现了当地干阑式民居的味道。驿站的建筑语言本身也具有当地古村落的特色，我们把建筑的屋顶设计得高低起伏，用了一些参数将屋顶折成不同的角度，抽象成独特的聚落形态，探索了乡土建筑的单元式转译。建筑采用钢木结构，主体采用钢结构，屋面采用与木结构结合的玻璃和小青瓦，再与当地传统的灰砖结合在一起，建筑的结构和材料体现出传统和现代的结合。

在施工的时候，这些屋顶交接的位置比较难处理。首先瓦片的铺砌没有那么简单，有些地方瓦要切掉半片，之后想办法固定。这些涉及施工方面的，都需要在现场解决。

做方案的时候，我们预留了村民继续发展建筑的可能性，在台阶上设置了座位，不过从后来村民使用的情况来看，他们更喜欢自己带着凳子过去坐，可能这是乡村的一个特点，使得建筑有更大的灵活性与包容性。

Meiling Station, a small project of only a few hundred square meters, is located near South China Historical Trail, a linear cultural heritage in Guangdong, in what we call today an economically underdeveloped area. The selected site for the project is in Nanxiong along Meiguan Ancient Road, which is a very famous tourist area in Guangdong.

The project nestles aside a large parking lot near the entrance to Meiling Ancient Village. During the selection of site location, we did have considered some other options, which, however, caused dissatisfaction from the villagers as some of their houses would have to be dismantled to make way. That's why we eventually selected the current site, which is right next to the local small shops and F&B, to minimize the impact on villagers' lives. Our consideration was highly recognized by the local and provincial government leaders, and welcomed by the villagers, as well.

The project is a simple two-storey building expected to provide multiple functions, serving tourists from the parking lot while creating space for villagers to rest and socialize. F1 is intended for ticketing and publicity. F2 has some semi-outdoor space for relaxation, where one can enjoy the mountain views, or for promotional activities, realizing very flexible functions. Clearly, what we intend to create is a grey space suitable for the Lingnan climate, which can accommodate the diversified activities of both tourists and villagers.

To highlight the local characteristics, we visited the area to explore the local building structures and details. Finally, some protruding balconies, supported by steel structures, were designed in the north of the Historical Trail, to resemble the typical local stilt houses. The architectural language of the station itself also demonstrates features of the ancient village, with undulating building

roofs folded into different angles and abstracted into unique settlement forms, an exploration of the modular interpretation of vernacular architecture. The building structure and materials also reflect the integration of tradition and modernity, with steel and wood for building structure, steel structure for the main building, glass, grey tiles and wood structure for the roof, and the use of traditional grey bricks.

Construction of buildings and roof connections posed some challenges. First, the installation of roof tiles was not that easy. Some had to be cut in half before they could be fixed securely, and all the construction-related problems encountered had to be solved right on site.

In our building design, we had reserved some possibility for our villagers, such as some seats on the steps. However, in the subsequent use, they simply preferred their own stools, probably another trait of the village, which also makes the building more flexible and inclusive.

广州猎桥桥西 110kV 变电站
110kV Lieqiao Substation, Guangzhou

项目地点：广州市，天河区
设计时间：2018- 2020 年
建设时间：2020- 2021 年
建筑面积：4,992 平方米
合作单位：广州电力设计院有限公司
建筑师团队：陈雄、黄俊华、高原、陈俊明、陈仁杰、陈细明、黄晨虹、杨竣凯、胡冰清、郑培鑫
主要奖项：电力行业（火电、送变电）优秀工程设计一等奖
广东省工程勘察设计行业协会科学技术奖一等奖
广东省土木工程詹天佑故乡杯奖

Location: Tianhe District, Guangzhou City
Design: 2018-2020
Construction: 2020-2021
GFA: 4,992m²
Partners: Guangzhou Power Design Institute Co., Ltd.
Design Team: Chen Xiong, Huang Junhua, Gao Yuan, Chen Junming, Chen Renjie, Chen Ximing, Huang Chenhong, Yang Junkai, Hu Bingqing, Zheng Peixin
Major Award(s): The First Prize (Power Industry - Thermal Power and Power Transmission and Distribution) of Excellent Engineering Design Award, The First Prize of Science and Technology Award of Guangdong Engineering Exploration & Design Association, Tien-yow Jeme Hometown Cup Prize by Guangdong Society of Civil Engineering and Architecture

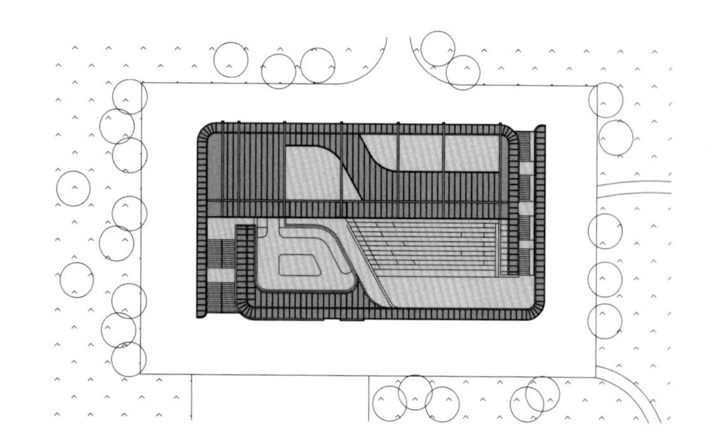

变电站是一个城市重要的基础设施，也是一种邻避设施。传统变电站有较大的噪声，可能还有辐射、污染等问题，难以得到市民的喜爱，所以我们希望能够对这种设施进行提升。

猎桥变电站位于广州新轴线和猎德桥的旁边，紧邻珠江，景观位置特别重要。当时规划局将这个变电站在内的六个项目作为第一批的“社区事 · 大师做”工程，让我们选了一个。变电站原来在桥的东面，但是大多居民都比较反对，政府协调了很多次，最后决定把变电站从桥东搬到桥西，更靠近商业设施，而不是居民住宅。所以从选址上来讲，应该说政府为了化解这些矛盾，也作了很好的安排，这是值得肯定的。

我们考虑这个项目除了满足变电站的功能，把外形做得比较漂亮之外，怎样可以创造更多的价值，最后提出了把这个点作为对城市开放的空间的核心思路，让这个工业建筑带有了公共属性。

我们首先将建筑朝向珠江，形成从珠江到岸边逐渐抬高的形态，另外我们想把屋顶利用起来，作为城市的小客厅，希望市民可以在这里面休息，欣赏珠江景观，也可以搞小型活动，设置一个表演场地，这样其实也很有意思。我们还与南方电网、广州供电局业主一起，把这个地方打造成电力展览教育基地，展示先进的电力技术。所以我们赋予了这座变电站“1+3”的定位，它的本体是变电站，同时还是展厅，也是城市小客厅，还是城市景观的一部分，探讨了变电站的新范式。

具体做的过程中，我们在屋顶花园东西两边各设有一部楼梯，这样可以形成环形流线；屋顶上做了太阳能光伏发电；场地里做了海绵技术，形成城市雨水回用，打造绿色三星工业建筑。

这个建筑有一个特点，整个立面使用了穿孔铝板的表皮。我们希望铝板的造型比较柔润、通透。同时解决一些功能开口，用于进风和排风。

我们注意到夜景的处理，因为整个片区很重要，小蛮腰、海心沙、琶洲、东塔、西塔都可以看到这座变电站，所以我们采用精心设计的灯光来打造建筑的夜景效果。我们反复调整灯光设计和播放内容，建筑的灯光有很多种模式，除了平日、节日模式外，还有很多丰富多彩的模式，调了很多遍。这需要编写动画，并结合建筑物的长度设置灯光的路径，利用 LED 灯技术，实现色彩的动态变化，使建筑的夜景非常丰富。

As an important urban infrastructure, substation is also a NIMBY facility that is always shunned by citizens. Yet for this project, we hope to make a difference.

Leiqiao Substation is planned besides Liede Bridge, the city's new central axis and the Pearl River, enjoying a strategic location for the landscape. The previous site east of the bridge was opposed by most residents, while through government coordination, the substation was relocated to the west of the bridge to get nearer to the commercial area, which indeed is a wise move.

Apart from the function and attractive appearance, we intend to create more values and thus propose the core concept of making the substation an urban open space, and an industrial building with public attributes.

First, the building is oriented toward the Pearl River with the building height ascending gradually from the river to the bank. Then the roof is used as a small city parlor, and, through cooperation with Southern Power Grid and Guangzhou Power Supply Bureau, the project is turned into a power exhibition and education base to communicate leading power technologies. In this way, the project realizes its "1+3" positioning as a substation, an exhibition facility, a small city parlor and a part of cityscape, exploring a new paradigm for substation design.

A staircase is provided respectively on the east and west of the roof garden to form a circular circulation. With solar photovoltaic power system on the roof and sponges at site for stormwater reuse, a three-star green industrial building was created.

The façade is fully clad with perforated aluminum panels. Meanwhile, functional openings can facilitate the air supply and exhaust.

The project area is quite important as the substation is visible from Canton Tower, Haixinsha Island, Pazhou Island, the CTF Finance Center and Guangzhou IFC. To achieve an impressive night view of the project, flood lighting is elaborately designed to ensure the desired lighting effect, contents and paths. LEDs lights are employed to bring dynamic color changes and fantastic night views to the building.

广州白云山柯子岭门岗
Baiyun Mt. Entrance (Keziling), Guangzhou

项目地点：广州市，白云区
设计时间：2019 年
建设时间：2019 年
建筑面积：1,565 平方米
建筑师团队：陈雄、黄俊华、郭其轶、许尧强、陈康桃、陈俊明、倪俍、赖锐敏、蔡雨珊

Location: Baiyun District, Guangzhou City
Design: 2019
Construction: 2019
GFA: 1,565m^2
Design Team: Chen Xiong, Huang Junhua, Guo Qiyi, Xu Yaoqiang, Chen Kangtao, Chen Junming, Ni Liang, Lai Ruimin, Cai Yushan

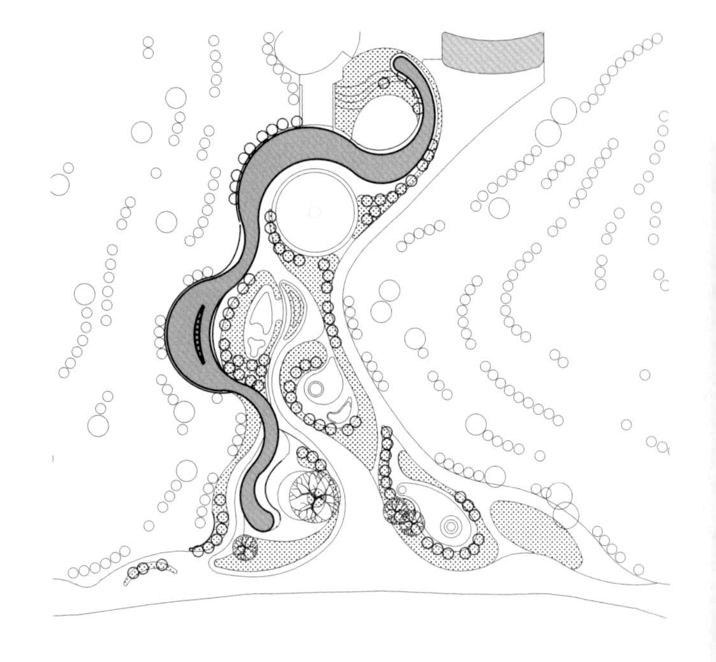

这个项目属于“社区事 · 大师做”工程系列，当时广州市要提升白云山的整体品质，需要把七个门岗重新设计提升。

这里原来是一个台地，我们将价值比较高的树都保留下来，一个很重要的特色就是这个门岗顺应了山势，很自然地做了入口，这与传统公园的大门完全不同。

在这个过程中，我们与园林局进行了很好的沟通，最早的时候入口在路边售票的地方，还设有一些临时停车场。第一，为了提升品质，我们与园林局协商，把停车场移到了一旁的空地上，使入口完全变成步行区；第二，我们将入口售票处挪到了里面，把大片的空地提供给周边市民，创造了一个免费的城市活动空间；第三，我们用一条灵动的游廊将游客服务中心、管理用房、售票处、公厕、电房等功能串联起来，市民可以在这里休息。

主入口上去后，又分为两条流线。右边设置无障碍设施，市民可以直接上到售票处，消防车也可以开上来。左侧游廊有意设置了台阶，形成与右边不同高度的活动平台。这些活动平台成了市民跳广场舞、打太极拳的场地。特别值得一提的，就是我们利用场地上这棵保留的大榕树做了一个榕树广场。最终整个游廊实现了“云山珠水”的云廊立意。

建好以后，大家看到航拍，都觉得像五羊雕塑。我们设计时其实没有这种想法，但却出现了这种有趣的巧合，我们的初心就是做一个好的城市公共空间。

很多铺地都是在现场处理的。连廊下灵动的地方也有高差，面对不同的台阶，我们把场地作为底图统一方向铺砌石材，上面做平台、绿地、台阶，整体性比较好，施工便捷。

关于连廊，还有一些施工方面的故事。柱头上的灯原来是扇面的，很难做出来，后来我看到一些商户店面的发光字，我们就叫施工队按照其工艺来做。柱子的底座是一个整体的扇形，使用了 GRG 人造石。

This project is one of the seven entrances to be redesigned and upgraded in an important green restoration initiative of Guangzhou to improve the overall quality of Baiyun Mountain, and one of the Masters' Community Design Works series projects.

Some precious trees are retained on the site, which was originally a tableland. An important feature of the entrance is its natural formation following the mountain terrain, which is completely different from the stereotyped park gate.

During project design, we made sufficient coordination with the garden bureau. Originally, the

entrance was positioned near the ticket office on the roadside, with a temporary parking lot. In order to improve the quality, firstly, we discussed with the garden bureau to relocate the parking lot to a vacant space nearby, hence transforming the entrance into a pedestrian area; secondly, we moved the ticket office inside from outside the entrance, leaving the open front area to the public, and in this way, creating a free urban activity space; thirdly, we connected the tourist service center, management facilities, ticket office, public toilets, electricity rooms etc. via a flexible corridor, where people can relax.

The main entrance then splits into two routes, with barrier-free facilities on the right for direct access to the ticket office by both citizens and the fire engine, and corridor steps on the left to create platforms for urban activities at different levels, such as square dancing and Tai Chi. A banyan tree square, in particular, is created centering around a large banyan tree retained on the site. Eventually, the design concept of Baiyun Mountain and Pearl River is realized.

The completed project looks like the "Five Rams" Statue in aerial photos, which is indeed an unexpected coincidence, as our original intention simply aimed at creating pleasant urban public spaces.

Most problems with the paving had to be resolved on site. Height differences exist even in the flexible spaces under the corridor. To tackle the step issues, we provided the site a uniform stone paving, and added platforms, green spaces and steps on it to ensure the consistency and facilitate the construction.

Corridor construction was not smooth, either. The light on top of the column was originally in a fan shape, which was difficult to realize. Later, inspired by the luminous storefront characters, we asked the construction team to deal with the light in a similar fashion. Besides, the column bases are also in a fan shape, and constructed with GRG artificial stones.

广州明珠湾起步区城市总师城市风貌管控

Urban Landscape Control of Guangzhou Mingzhu Bay Kick-off Zone under Chief Urban Designer Mechanism, Guangzhou

项目地点：广州市，南沙区
设计时间：2021 年
用地面积：12.3 公顷
建筑师、规划师团队：
陈雄、洪卫、李鹏、任小蔚、区绮雯、刘小丽、邓欣、李珏、甘慧盈、冯财坚、朱佳怡、杜嘉浩、王妍、罗玮、蔡晓辉、马天天、吴嘉俊、梁宇明、张宛馨

Location: Nansha District, Guangzhou City
Design: 2021
Land Area: 12.3hm²
Design Team: Chen Xiong, Hong Wei, Li Peng, Ren Xiaowei, Ou Qiwen, Liu Xiaoli, Deng Xin, Li Jue, Gan Huiying, Feng Caijian, Zhu Jiayi, Du Jiahao, Wang Yan, Luo Wei, Cai Xiaohui, Ma Tiantian, Wu Jiajun, Liang Yuming, Zhang Wanxin

这个城市设计项目位于明珠湾起步区西南部的横沥岛尖上，是一个具有自然禀赋的区域，有很丰富的水系与珠江连接在一起，陆地形态比较生动。为了提升横沥岛的综合价值，我们根据《横沥岛尖规划建设评估》文件，开展了横沥岛尖城市设计的优化编制工作，从项目的发展定位、规划系统设计、城市空间特色等不同层面优化了原来的方案。

这个项目有三个主要的组团，包括国际金融岛组团、科技产业组团和高端配套组团。我作为城市片区的总规划师，与三个总师进行配合，把高端配套服务组团作为我们管控的工作范围，建立了全周期、全方位的规划、设计、建设、实施的技术服务咨询平台，通过多个专业团队的参与，为南沙新区明珠湾开发建设管理局管理工作，以及广州市规划和自然资源局南沙区分局行政审批提供依据，为明珠湾起步区的开发建设提供了有力的技术支持，实现精细化管理，协调指导各类规划、工程的落地实施。

我们在项目中提出了高效活力、回归水域、幸福家园、快城慢里的优化理念。地块中间有一条中轴涌，两边还有靠近珠江的滨江景观资源。我们希望能够把中轴涌的价值发挥出来，就围绕中轴涌做了一些滨水的商业休闲设施。人们在这里可以进行亲水活动，有点像新加坡滨水的商业休闲设施，也有一些水景的设计。另外，我们将这上面的两个轨道周边的地块做了一个 TOD 的开发。从整个城市设计方面，我们进行了用地、景观和公共服务设施的规划，形成全方位、多层次的完善功能。岛尖是 IFF 国际金融论坛的场所。

地区总设计师的城市风貌管控的工作特色，一是以守住底线，彰显特色作为风貌管控的原则；二是以人为本，在全过程中保障公众利益；三是用伴随式服务帮助地区的发展建设。这个项目从评估工作、城市设计优化到地区总师管控，经历了将近三年的时间，面临着多重的挑战，项目时间紧、复杂程度高、难度挑战大。

This urban design project is located at the tip of Hengli Island in the southwest of Mingzhu Bay Kick-off Zone. The region is blessed with natural endowments, such as the abundant water systems connected to the Pearl River, and vivid land forms. In order to maximize the integrated value of Hengli Island, we have carried out urban design optimization according to *Hengli Island Tip Planning and Construction Evaluation*, upgrading the original design in different aspects from development positioning, planning system design to urban space features.

This project consists of three main clusters, namely, the International Financial Island Cluster, the Technology Industry Cluster and the High-end Supporting Service Cluster. As chief urban architect coordinating with the other two chief architects, I focused on our main work scope of the high-end supporting service cluster and established a technical service consulting platform for full-cycle and all-round planning, design, construction and implementation. With the participation of several professional teams, the platform not only serves as the basis for the management of Pearl Bay Development Authority of Guangzhou Nansha New Area and the administrative review and approval of Nansha District Branch of Guangzhou Municipal Planning and Natural Resources Bureau, but also provides strong technical support for the development and construction of Mingzhu Bay Kick-off Zone, realizing fine management while coordinating and guiding the implementation of various plans and projects.

We proposed the optimization concept of efficiency and vitality, return to water, happy homeland, quick city with slow lanes for the project. With a central-axis river running through the site and riverfront landscape resources of the Pearl River on both sides, some waterfront commercial leisure facilities are designed along the central-axis river to maximize the value of the river, and provide space for people to get close to water, kind of like the waterfront commercial leisure facilities in Singapore, dotted with some water features. Besides, a TOD development plan is proposed for the surrounding area of the two rail transits above. In the overall urban design, we focused on the planning of land use, landscape and public service facilities and realized all-round and multi-level functions. The island tip is also the permanent site of International Finance Forum (IFF).

The responsibility of chief architect in urban landscape control is to firstly, hold the bottom line and highlight the principles for landscape control; secondly, put people first and protect the public interest throughout the entire process; thirdly, provide accompanying services for the development and construction of the region. From project evaluation, urban design optimization to regional control by the chief architect, which lasted nearly three years, the project has encountered multiple challenges, such as a tight schedule, high complexity and great difficulty.

广州荔湾海南经济联社更新城市设计

Urban Design for Renewal of Hainan Economic Association, Liwan, Guangzhou

项目地点：广州市，荔湾区
设计时间：2021 年
用地面积：483.6 公顷
建筑师、规划师团队：
陈雄、李鹏、任小蔚、区绮雯、李珏、王伟阳、甘慧盈、王妍、刘雅琦、刘玥雅、回雪静、张子洋、马千程、赵丽雅、徐文韬

Location: Liwan District, Guangzhou City
Design: 2021
Land Area: 483.6hm^2
Design Team: Chen Xiong, Li Peng, Ren Xiaowei, Ou Qiwen, Li Jue, Wang Weiyang, Gan Huiying, Wang Yan, Liu Yaqi, Liu Yueya, Hui Xuejing, Zhang Ziyang, Ma Qiancheng, Zhao Liya, Xu Wentao

海南村地处广佛交界的心脏地带，位于“广佛高质量发展融合试验区先导区”的核心位置。作为历史悠久的千年岭南水乡及荔湾花卉亿元村，留存多个宗祠、生菜会、龙舟会等历史文化资源，见证了广州水陆商贸的繁荣兴盛，是近现代花卉、制造产业变迁的重要见证。随着时代变迁、发展理念转变，海南村面临的生态空间受产业挤压、环境品质退化、历史人文辨识度式微等诸多问题。

为加快实现老城市新活力，海南村借助广佛同城建设发展契机，率先探索“大师做”城市更新 4.0 模式，全面参与旧村规划建设，引导后续高质量实施、释放存量资源新价值，引领“广佛边缘”走向“广佛同城核心”，打造广佛高质量融合发展示范城区。

结合城市设计与旧改更新，在开发量上“以新换旧”、功能上“以新带旧”、空间上“以新融旧”，聚焦生态、品质、特色、文化等综合效益的平衡，实现从用地到空间的价值最大化，推动海南村的产业经济和人居环境双重提升，延续与彰显海南的村落人文特色，构建“河心复源、花地半岛”的目标愿景。

规划联动海南村及周边，以 TOD 站点为触媒，以花地河及蓝绿格局为脉络，塑造疏密有致的“花水田城相融”的城市空间。形成四大设计特色：一是平衡高强度开发与生态资源保护，通过综合治理农用地与修复污染、引入现代农业科技、花卉上楼等手段提升农田生态价值、实现农业生产智能化。二是识别、保护和利用“水花田共生”的村落空间特色以及花卉、龙舟等文化符号，融合到海南村主题公共空间设计中，实现水脉、文脉相融合。三是依托临田、近水重要界面、多层次公共空间等要素，打造促交往、亲近邻、享配套的多元生活场景；四是探索“大师做”全流程精细化的工作模式。

Hainan Village is strategically located at the heart of the border between Guangzhou and Foshan, and the core of the pilot area for high quality development and integration of the two cities. As a millennium-old Lingnan-style water village, it has generated an annual income of 100 million yuan over its thriving flower industry. Meanwhile, it is home to diverse historical and cultural resources, including ancestral temples, lettuce fair, dragon boat fair. It has witnessed the prosperous commerce of Guangzhou, both on water and land, and the evolution of modern flower and manufacturing industries. Yet due to the changes of era and development concept, Hainan Village now faces various challenges, such as the ecological space encroached by industries, degraded environmental quality, and low historical and cultural visibility.

To support the revitalization of old urban areas in Guangzhou, Hainan Village seized the opportunity of Guangzhou-Foshan integrated development and took the lead in exploring the urban renewal version 4.0 "led by design masters". It has exerted all-around efforts in old village planning and redevelopment to guide the high-quality implementation at subsequent phase, unleash the potential value of available resources, lead the transformation from "Guangzhou-Foshan border" to "Guangzhou-Foshan integration core", and eventually create a demonstrative urban area for high-quality integrated development of the two cities.

Combining urban design with old village redevelopment and renewal, the project, by providing new development, functions and spaces, ensures a balanced overall benefits in terms of ecology, quality, characteristics, and culture. In doing so, it can maximize the value of land and space, improve the village's industrial economy and living environment, sustain and highlight the cultural characteristics of the Village and eventually realize the the vision of a flowering peninsula in river center.

Connecting the village with its surrounding area, the planning aims to create well-organized urban spaces that integrate flowers, water areas, farmlands and urban fabrics, with TOD stations as catalysts and Huadi River and the river/green spaces as framework. Accordingly, four design features are established. First, balance the high-intensity development and ecological resource protection to improve the ecological value of farmland and realize the intelligent agriculture through various means, such as comprehensive management of agricultural land and pollution remediation, use of modern agricultural technology, promoting flower industry in high buildings, etc; second, identify, protect and use village spaces featuring symbiosis of waters, flowers and farmlands and cultural symbols such as flowers and dragon boats, so as to incorporate them into the design of themed public space in the village and realize the integration of water features and cultural context; third, build diversified life scenes at interfaces along farmlands and waters, and multi-level public spaces and promote social contact, neighborhood relationship with full-fledged supporting facilities. Forth, explore the refined whole-process working mode "led by design masters".

广州荔湾海北村更新片区城市设计

Urban Design for Renewal of Haibei Village, Liwan, Guangzhou

项目地点：广州市，荔湾区
设计时间：2021 年
用地面积：112.66 公顷
建筑师、规划师团队：
陈雄、李鹏、任小蔚、区绮雯、李珏、王伟阳、甘慧盈、王妍、刘雅琦、刘玥雅、回雪静、张子洋、马千程、赵丽雅、徐文韬

Location: Liwan District, Guangzhou City
Design: 2021
Land Area: 112.66hm^2
Design Team: Chen Xiong, Li Peng, Ren Xiaowei, Ou Qiwen, Li Jue, Wang Weiyang, Gan Huiying, Wang Yan, Liu Yaqi, Liu Yueya, Hui Xuejing, Zhang Ziyang, Ma Qiancheng, Zhao Liya, Xu Wenyao

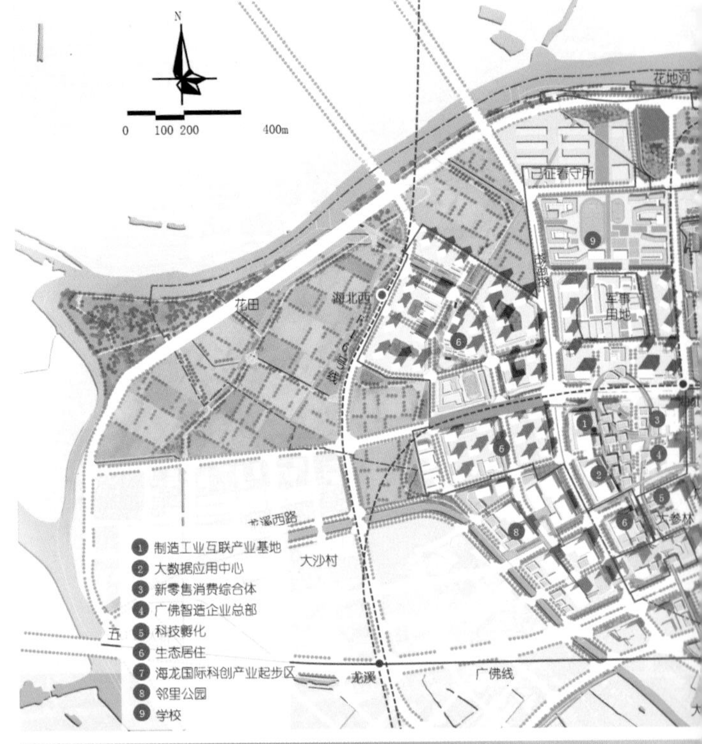

广佛自古是同根同源的关系。海北村位于广佛交界的心脏地带，是广佛经济、生活、文化高度交融的有机体；是非城非乡、功能混杂交错、用地低效无序的城市边缘；是白鹅潭、千灯湖、三龙湾等三强心间的价值洼地。怎样从发展逻辑、空间框架上挖掘地区价值，珍视农田、水乡等大自然的馈赠、传承古村、民俗等老祖宗的财富，是海北村转型发展的关键。

我们提出“桃源胜境、海北新生”的设计愿景，以蓝绿渗透、创享核芯、文化畅游、复合街坊四大规划策略构建“湾区智造芯 · 广佛宜居城”。提出了三大规划目标。一是发展路径从“粗放扩张”向“生态文明、有机增长”跨越，助力“双碳”目标实现。二是同城模式从“单项合作”向“人文导向、全面融合”跨越。探索有机更新、价值共享的深度融合。三是产业发展从“自发零散”向“赋能联动”跨越。

海北村将成为城与乡、自然与人文、创新与产业的耦合发生地，集科技创新、总部经济、电子商务、文化旅游、居住生活等多功能于一体的活力区域，成为宜居宜业宜游商务高品质人居环境的示范、广佛融合先导区的活力枢纽。

规划形成了五大设计特色：一是从需求出发，探索双城边界以产业为导向的城市空间设计。二是在守护中发展，构建生态资源高质量转化的新模式。三是基于岭南水乡与村落肌理特色探索兼顾开发价值与友好尺度的街区开发。四是多维价值叠加，促进滨水人文空间的活化新生，塑造沉浸式文旅新体验。五是以适度超前的城市设计为平台，打下良好的品质空间导控基础。

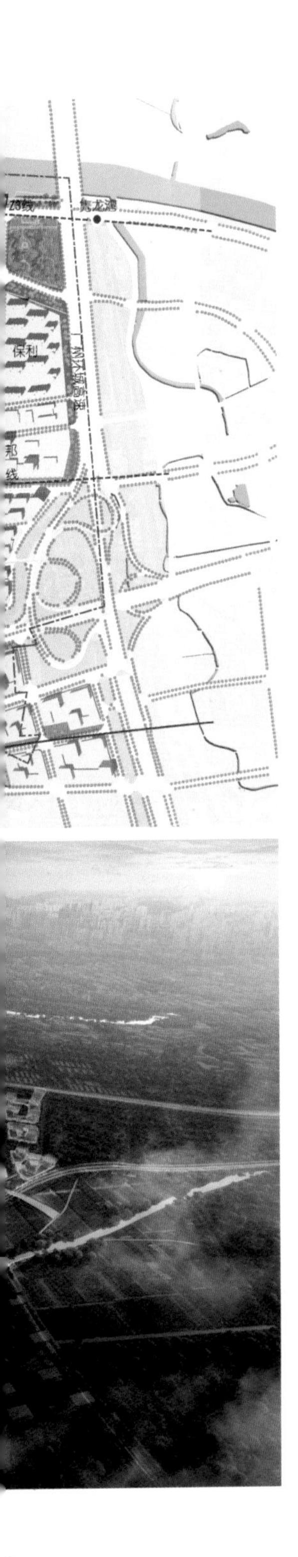

Located at the heart of the border between Guangzhou and Foshan, Haibei Village is an organic entity with highly integrated economy, life and culture of the two cities. As an urban fringe that is neither urban nor rural, it features interwoven, mixed functions and inefficient and unregulated development. It has been undervalued by the shadow of three development focuses, i.e. Bai'e Tan, Qiandeng Lake, and Sanlong Bay. Therefore the keys to the transformation and development of Haibei Village are to explore value from the development logic and spatial framework, value farmlands and water towns endowed by nature, and inherit the legacy of the ancient village and folk customs.

With the vision of "revitalizing Haibei Village into a land of idyllic beauty", the planning proposes four planning strategies, i.e. interwoven greenery and water areas, innovation core, cultural tourism, and mixed-use neighbourhood, to create an intelligent manufacturing core in the GBA and a livable city in Guangzhou-Foshan region. It also establishes three objectives. First, the development path shifts from "extensive expansion" to "environment-friendly and organic growth" in a bid to support the carbon peaking and carbon neutrality goals; second, the city integration changes from "cooperation on single project" to "people-oriented and all-round integration" to explore the in-depth integration featuring organic renewal and value sharing; third, industrial development leaps from the spontaneous and fragmented mode to empowered and coordinated development.

Haibei Village will become a coupling place that integrates city with village, nature with culture, and innovation with industry; a dynamic area for S&T innovation, headquarters economy, e-commerce, cultural tourism, residential life and other functions; an example for high-quality living environment that is friendly to life, work, tourism and business; and a dynamic hub in the pilot area.

The planning reflects five highlights. First, explore the demand-led, industry-oriented urban space design at the city border area; second, seek development in the process of conservation and develop a new mode for high-quality transformation of ecological resources; third, based on the fabrics of Lingnan water towns and villages, explore the neighborhood development that reflects both development values and friendly scales; fourth, superpose values of different dimensions to revitalize waterfront cultural spaces and create immersive cultural tourism experience; fifth, employ moderately forward-looking urban design as the platform to serve as effective guidance and regulation of quality spaces.

年表

BIOGRAPHY

1979-1986

1979 年进入华南工学院建筑学系学习建筑学，1983 年大学本科毕业后在华南工学院建筑设计研究院继续攻读硕士研究生，师从林克明、郑鹏两位教授，1986 年硕士研究生毕业。

1986-1998

1986 年进入广东省建筑设计研究院工作，从跟随前辈们学习实践到逐步独立担纲。1998 年带领设计小组参加广州白云国际机场一号航站楼的国际竞赛，虽未中标，但由于体现了较好的技术水平，被业主选定为与国外公司合作共同设计新航站楼，开始职业生涯的重要阶段。

1998-2010

1998 年起与国外公司展开广州白云国际机场一号航站楼项目合作，积累了丰富的大型复杂公共建筑（特别是大跨度大空间建筑）设计经验。2004 年成立广东省建筑设计研究院机场设计研究所，有了相对固定的团队。从交通建筑延伸到体育建筑，团队以中国建筑师原创方案获胜并实施了 2010 年广州亚运会主场馆。

2010-2022

以机场航站楼和体育场馆两种主要项目类型为主轴，不断深入研究和实践，同时项目类型向更加多元化发展，包括会展类、开发类等大型公共建筑，也包含了中小型的文化建筑、工业建筑，并持续关注城市设计和乡村振兴。担任中国建筑学会常务理事、中国民航工程咨询专家、广东省工程勘察设计行业协会建筑专业委员会主任委员、广东省土木建筑学会常务理事、广东省注册建筑师协会常务理事、广州市城市规划委员会委员等社会兼职，推动行业发展和技术进步。

提出“好设计，用心做”的团队核心理念。

1979-1986 Studying architecture in South China Institute of Technology (SCIT, now South China University of Technology, SCUT) since 1979 and earning a bachelor's degree in 1983, Chen Xiong continued to pursue his master degree study in SCIT Architectural Design Institute under the supervision of professors Lin Keming and Zheng Peng, and graduated with a master's degree in 1986.

1986-1998 Cheng Xiong started his career as an architect at GDADRI in 1986. By learning from his superiors, he kept improving his expertise and gradually became a competent architect to work independently on projects. In 1998, he led a design team to participate in the international bidding for Terminal 1 of Guangzhou Baiyun International Airport. Though the team was not selected as the bid winner, the client was much impressed by their expertise and decided to engage the team as the local design office to work with the foreign designer on the new terminal, thereon opening an important chapter in Chen's career.

1998-2010 From 1999, Chen Xiong and his team began to work with the foreign architects on Terminal 1 of Guangzhou Baiyun International Airport, and gained enormous design experience on large-scale complex public buildings (in particular those with large span and large space). In 2004, he built a relatively stable team by establishing the Airport Design and Research Institute of GDADRI. The team expanded their efforts from transportation buildings to sports buildings, and, with their own creative design scheme, won the design contract of and successfully implemented the Main Venue for the 2010 Guangzhou Asian Games.

2010-2022 While focusing on two project typologies, i.e. airport terminals and sports venues, Chen Xiong led his team to work on a wider range of projects, from large-scale public buildings for convention and exhibition, to small- and medium-sized ones for cultural and industrial purposes. He also had a sustained interest in rural revitalization. He has been keen in promoting the advancement of the design industry and technology through professional affiliations. He has served as the Executive Board Member of Architectural Society of China, Engineering Consulting Expert of Civil Aviation Administration of China, and Director of the Architectural Professional Committee of Guangdong Engineering Survey & Design Association.

Delivering quality design with due diligence has remained the core philosophy of his team.

重要记事

MILESTONES

- 1962 年 出生于广东广州
- 1979 年 入读华南工学院（现华南理工大学）建筑学专业
- 1983 年 本科毕业，获学士学位
- 1983 年 入读华南工学院（现华南理工大学）建筑设计研究院建筑设计专业研究生
- 1986 年 研究生毕业，获硕士学位
- 1986 年 就职于广东省建筑设计研究院
- 1986- 1988 年 任广东省建筑设计研究院三室建筑师
- 1988- 1995 年 任广东省建筑设计研究院三室 / 三所高级建筑师
- 1995- 1996 年 任广东省建筑设计研究院策划室主任兼总建筑师
- 1996- 1998 年 任广东省建筑设计研究院二所副总建筑师、总建筑师
- 1999- 2004 年 任广东省建筑设计研究院副总建筑师
- 2004- 2014 年 任广东省建筑设计研究院副总建筑师兼机场设计研究所所长
- 2013 年 任广东省建筑设计研究院 ADG 建筑创作工作室主任
- 2014- 2015 年 任广东省建筑设计研究院总建筑师
- 2015- 2020 年 任广东省建筑设计研究院副院长
- 2020- 2022 年 任广东省建筑设计研究院有限公司董事、副总经理
- 2021 年至今 任广东省建筑设计研究院有限公司首席总建筑师

- 1998 年 获国家一级注册建筑师资格
- 2002 年 获教授级高级建筑师职称
- 2004 年 获“广东省建设系统先进工作者”称号
- 2007 年 获“广东省五一劳动奖章获得者”称号
- 2007 年 获“全国建设系统先进工作者”称号
- 2012 年 获“当代中国百名建筑师”称号
- 2016 年 获住房和城乡建设部颁发“全国工程勘察设计大师”称号及证书
- 2016 年 获“享受国务院特殊津贴专家”荣誉
- 2019 年 获全国勘察设计同业协会“杰出人物”称号
- 2019 年 获“建国 70 周年暨第一届中国建筑设计行业管理卓越人物 · 最佳突出贡献奖”称号

- 1962 Born in Guangzhou, Guangdong Province
- 1979 Studied architecture at South China Institute of Technology (SCIT, now SCUT)
- 1983 Graduated with a bachelor's degree
- 1983 Pursued postgraduate studies in architectural design at SCIT (now SCUT)
- 1986 Graduated with a master's degree
- 1986 Joined Guangdong Architectural Design and Research Institute (GDADRI, Now GDAD)
- 1986-1988 Worked as architect with the 3rd Studio, GDADRI
- 1988-1995 Acted as Senior Architect of the 3rd Studio / the Third Branch Institute, GDADRI
- 1995-1996 Acted as Director and Chief Architect of the Planning and Organization Studio, GDADRI
- 1996-1998 Acted as Deputy Chief Architect/Chief Architect of the Second Branch Institute, GDADRI
- 1999-2004 Acted as Deputy Chief Architect of GDADRI
- 2004-2014 Acted as Deputy Chief Architect of GDADRI & Head of Airport Design and Research Institute of GDADRI
- 2013 Acted as Director of the ADG Architectural Creation Studio, GDADRI
- 2014-2015 Acted as Chief Architect of GDADRI
- 2015-2020 Acted as Vice President of GDADRI
- 2020-2022 Acted as Director & Deputy General Manager of GDAD
- 2021 to present Acted as Chief Architect of GDAD

- 1998 Certified as a First-class Registered Architect of China
- 2002 Awarded the title of Professor-level Senior Architect
- 2004 Awarded the title of Advanced Worker of Guangdong Construction Sector
- 2007 Awarded Guangdong May 1st Labor Medal
- 2007 Awarded the title of Advanced Worker of the National Construction Sector
- 2012 Enlisted as one of the 100 Contemporary Chinese Architects
- 2016 Awarded the title and certificate of Engineering Survey and Design Master of China by MoHURD
- 2016 Honored as an Expert Entitled to the State Council Special Allowance
- 2019 Awarded the title of Outstanding Individual by China Engineering and Consulting Association (CECA)
- 2019 Awarded the Most Outstanding Contribution Award for the 70th Founding Anniversary of the People's Republic of China and the First Session of Excellent Management Figure Selection of Chinese Architectural Design Industry

主要获奖作品

- **广州白云国际机场一号航站楼**
  全国优秀工程设计金奖
  詹天佑土木工程大奖
  “全国十大建设科技成就”称号
  全国绿色建筑创新奖
  中国“百年百项杰出土木工程”
  中国建筑学会建筑创作大奖
  广东省优秀工程设计一等奖
  广东省优秀工程技术创新奖

- **东莞中心区海德广场**
  全国优秀工程勘察设计奖 · 建筑工程二等奖
  香港建筑师学会两岸四地建筑设计大奖卓越奖
  广东省优秀工程勘察设计奖 · 工程设计一等奖

- **武汉火车站（参与设计）**
  芝加哥雅典娜建筑设计博物馆颁发“国际建筑奖”
  中国土木工程詹天佑奖
  中国“百年百项杰出土木工程”
  铁路优质工程勘察设计一等奖

- **惠州金山湖游泳跳水馆**
  国家优质工程银质奖
  全国优秀工程勘察设计行业奖 · 建筑工程二等奖
  中国建筑学会建筑创作佳作奖
  广东省优秀工程勘察设计奖 · 工程设计二等奖
  广东省注册建筑师协会优秀建筑佳作奖

- **广州科学城科技人员公寓**
  全国优秀工程勘察设计奖住宅与住宅小区一等奖
  中国建筑学会建筑创作优秀奖
  广东省优秀工程勘察设计奖工程设计二等奖
  香港建筑师学会两岸四地建筑设计论坛及大奖住宅类卓越奖
  广东省注册建筑师协会第六次优秀建筑创作奖

- **揭阳潮汕国际机场航站楼**
  全国优秀工程勘察设计行业奖 · 建筑工程公建三等奖
  香港建筑师学会两岸四地建筑设计大奖卓越奖
  广东省优秀工程勘察设计奖 · 工程设计二等奖

SELECTED AWARD-WINNING PROJECTS

- **Terminal 1, Guangzhou Baiyun International Airport, Guangzhou**
  The Gold Award of National Excellent Engineering Design Award
  Tien-yow Jeme Civil Engineering Prize
  The Top Ten Construction Technological Achievements in China
  The National Green Building Innovation Award
  100 Outstanding Civil Engineering Projects from 1900 to 2010 by China Civil Engineering Society (CCES)
  ASC Architectural Creation Award
  The First Prize of Excellent Engineering Design Award of Guangdong Province
  Excellent Engineering Technology Innovation Award of Guangdong Province

- **Dongguan Commercial Center (Haide Plaza), Dongguan**
  The Second Prize (Construction Engineering) of National Excellent Engineering Exploration and Design Award
  Nominated Award of Hong Kong Institute of Architects (HKIA) Cross-Strait Architectural Design Symposium and Awards
  The First Prize (Engineering Design) of Excellent Engineering Exploration and Design Award of Guangdong Province

- **Wuhan Railway Station, Wuhan (as an architect)**
  The Chicago Athenaeum Museum of Architecture and Design "International Architecture Award"
  Tien-yow Jeme Civil Engineering Prize
  100 Outstanding Civil Engineering Projects from 1900 to 2010 by China Civil Engineering Society
  The First Prize of Excellent Railway Engineering Exploration and Design Award

- **Jinshan Lake Swimming and Diving Natatorium, Huizhou**
  National Quality Engineering Sliver Award
  The Second Prize (Construction Engineering) of National Excellent Engineering Exploration and Design Industry Award
  Honorable Mention of the ASC Architectural Creation Award
  The Second Prize (Engineering Design) of Excellent Engineering Exploration and Design Award of Guangdong Province
  Honorable Mention of the Excellent Architectural Creation Award of Guangdong Province Registered Architect Association

- **Guangzhou Science City S & T Talent Apartments, Guangzhou**
  National Quality Engineering Sliver Award
  The Second Prize (Construction Engineering) of National Excellent Engineering Exploration and Design Industry Award
  Honorable Mention of the ASC Architectural Creation Award
  The Second Prize (Engineering Design) of Excellent Engineering Exploration and Design Award of Guangdong Province
  Honorable Mention of the Excellent Architectural Creation Award of Guangdong Province Registered Architect Association

- **Jieyang Chaoshan International Airport Terminal, Jieyang**
  The Third Prize (Construction Engineering - Public Building) of National Excellent Engineering Exploration and Design Industry Award
  Nominated Award of Hong Kong Institute of Architects (HKIA) Cross-Strait Architectural Design Symposium and Awards
  The Second Prize (Engineering Design) of Excellent Engineering Exploration and Design Award of Guangdong Province

- **广州亚运馆**
  AAA 亚洲建筑师协会荣誉奖
  中国“百年百项杰出土木工程”
  全国优秀工程勘察设计行业奖 · 建筑工程一等奖
  绿色建筑与低能耗建筑“双百”示范工程
  中国建筑学会建筑创作优秀奖
  中国土木工程詹天佑奖
  詹天佑土木工程大奖 · 创新集体奖
  中国室内设计年度优秀公共空间设计金堂奖
  广东省优秀工程勘察设计奖 · 工程设计一等奖
  广东省注册建筑师协会优秀建筑创作奖
  广东省土木工程詹天佑故乡杯奖
  香港建筑师学会两岸四地建筑设计大奖优异奖

- **广州花都东风体育馆**
  全国优秀工程勘察设计行业奖 · 建筑工程三等奖
  中国室内设计大奖赛学会奖
  广东省优秀工程设计二等奖

- **深圳宝安国际机场新航站区交通中心**
  全国优秀工程勘察设计行业奖 · 建筑工程二等奖
  广东省优秀工程勘察设计奖二等奖

- **广州白云国际机场二号航站楼**
  SKYTRAX“全球五星航站楼”认证
  SKYTRAX“全球最杰出进步机场”
  SKYTRAX“中国最佳机场员工”
  CAPSE 全球最佳机场
  WBIM 国际数字化大奖
  国家三星级绿色建筑设计标识证书
  国家三星级绿色建筑运行标识证书
  全国绿色建筑创新奖一等奖
  行业优秀勘察设计奖 · 优秀（公共）建筑设计一等奖
  行业优秀勘察设计奖 · 优秀绿色建筑一等奖
  中国建筑学会建筑设计奖 · 公共建筑一等奖
  中国建筑学会建筑设计奖 · 幕墙技术一等奖
  中国威海国际建筑设计大奖赛银奖
  广东省优秀工程勘察设计奖 · 建筑工程一等奖
  广东省优秀工程勘察设计奖 · 建筑装饰一等奖
  广东省优秀工程勘察设计奖 · 园林和景观工程二等奖
  广东省优秀工程勘察设计奖 · BIM 专项一等奖
  广东省土木工程詹天佑故乡杯奖

- **Guangzhou Asian Games Gymnasium, Guangzhou**
  ARCASIA Awards for Architecture (AAA)
  100 Outstanding Civil Engineering Projects from 1900 to 2010 by China Civil Engineering Society
  The First Prize (Construction Engineering) of National Excellent Engineering Exploration and Design Industry Award
  One Hundred Green Building/Low-energy Building Demonstration Project
  Excellence Award of the ASC Architectural Creation Award
  Tien-yow Jeme Civil Engineering Prize
  Tien-yow Jeme Civil Engineering Prize (Collective Innovation)
  Jintang Prize-China Interior Design Award
  The First Prize (Engineering Design) of Excellent Engineering Exploration and Design Award of Guangdong Province
  Excellent Architectural Creation Award of Guangdong Province Registered Architect Association
  Tien-yow Jeme Hometown Cup Prize by Guangdong Society of Civil Engineering and Architecture
  Merit Award of Hong Kong Institute of Architects (HKIA) Cross-Strait Architectural Symposium and Design Awards

- **Dongfeng Gymnasium, Huadu District, Guangzhou**
  The Third Prize (Construction Engineering) of National Excellent Engineering Exploration and Design Industry Award
  China Institute of Interior Design (CIID) China Interior Design Awards
  The Second Prize of Excellent Engineering Design Award of Guangdong Province

- **Shenzhen International Airport New Terminal Area Transportation Center**
  The Second Prize (Construction Engineering) of National Excellent Engineering Exploration and Design Industry Award
  The Second Prize of Excellent Engineering Exploration and Design Award of Guangdong Province

- **Terminal 2, Guangzhou Baiyun International Airport, Guangzhou**
  SKYTRAX "Global Five-star Terminal"
  SKYTRAX "The World's Most Improved Airport"
  SKYTRAX "The Best Airport Staff in China"
  CAPSE The World's Best Airport
  WBIM International Digitalization Award
  China Green Building Label (Three-star)
  China Green Building Operation Label (Three-star)
  The First Prize of National Green Building Innovation Award
  The First Prize of Excellent (Public) Architecture Design of National Excellent Exploration and Design Industry Award
  The First Prize of Excellent Green Building Design of National Excellent Exploration and Design Industry Award
  The First Prize (Public Building) of ASC Architectural Design Award
  The First Prize (Facade Technology) of ASC Architectural Design Award
  Sliver Award of the Weihai International Architectural Design Grand Prix
  The First Prize (Construction Engineering) of Excellent Engineering Exploration and Design Award of Guangdong Province
  The First Prize (Building Decoration) of Excellent Engineering Exploration and Design Award of Guangdong Province
  The Second Prize (Landscape Architecture) of Excellent Engineering Exploration and Design Award of Guangdong Province
  The First Prize (BIM) of Excellent Engineering Exploration and Design Award of Guangdong Province
  Tien-yow Jeme Hometown Cup Prize by Guangdong Society of Civil Engineering and Architecture

- **香港新福港地产 · 佛山新福港广场**
  行业优秀勘察设计奖 · 优秀住宅与住宅小区设计一等奖
  全国 BIM 大赛（施工组）二等奖
  广东省优秀工程勘察设计奖 · 住宅与住宅小区二等奖

- **珠海横琴保利中心**
  世界建筑师大会（UIA）中国馆入选作品
  国家三星级绿色建筑设计标识证书
  行业优秀勘察设计优秀奖 · （公共）建筑设计一等奖
  行业优秀勘察设计优秀奖 · 绿色建筑设计二等奖
  国家城市设计试点珠海优秀项目
  中国建筑学会建筑设计奖 · 公共建筑三等奖
  中国建筑学会建筑设计奖 · 绿色生态技术一等奖
  中国威海国际建筑设计大奖赛银奖
  广东省优秀工程勘察设计奖 · 建筑工程一等奖
  广东省优秀工程勘察设计奖 · BIM 专项二等奖
  广东省优秀工程勘察设计奖 · 绿色建筑工程设计二等奖
  广东省注册建筑师协会优秀建筑佳作奖
  广东省注册建筑师协会广东省建筑设计奖 · 建筑方案奖公建类二等奖
  广东省土木工程詹天佑故乡杯奖
  广东省 BIM 应用大赛一等奖
  香港建筑师学会两岸四地建筑设计大奖 · 商业类银奖（金奖空缺）

- **肇庆新区体育中心**
  行业优秀勘察设计奖 · 优秀（公共）建筑设计一等奖
  中国建筑学会建筑设计奖 · 公共建筑三等奖
  中国威海国际建筑设计大奖赛铜奖
  广东省优秀工程勘察设计奖 · 建筑工程一等奖
  广东省优秀工程勘察设计奖 · 绿色建筑专项二等奖
  广东省土木工程詹天佑故乡杯奖
  广东省土木建筑学会科学技术奖

- **深圳宝安国际机场卫星厅**
  SKYTRAX“全球五星航站楼”认证
  CAPSE 全球最佳机场
  国际奖项协会缪斯设计奖铂金奖
  广东省绿色建筑评价标识 · 三星级绿色建筑设计标识证书
  广东省优秀工程勘察设计奖 · 建筑信息模型（BIM）专项二等奖
  广东省注册建筑师协会广东省建筑设计奖 · 建筑方案奖公建类一等奖

- **SFK Properties · SFK Plaza, Foshan**
  The First Prize (Excellent Residential Building and Quarter) of National Excellent Exploration and Design Industry Award
  The Second Prize of the National BIM Competition (Construction)
  The Second Prize (Residential Building and Quarter) of Excellent Engineering Exploration and Design Award of Guangdong Province

- **The Poly Center in Hengqin, Zhuhai**
  Shortlist of China Pavilion, the International Union of Architects (UIA) World Congress of Architect
  China Green Building Label (Three-star)
  The First Prize of Excellent (Public) Architecture Design of National Excellent Exploration and Design Industry Award
  The Second Prize of Excellent Green Building Design of National Excellent Exploration and Design Industry Award
  Excellent Project Award of National Urban Design Pilot City in Zhuhai
  The Third Prize (Public Building) of ASC Architectural Design Award
  The First Prize (Green Ecological Technology) of ASC Architectural Design Award
  Sliver Award of the Weihai International Architectural Design Grand Prix
  The First Prize (Construction Engineering) of Excellent Engineering Exploration and Design Award of Guangdong Province
  The Second Prize (Green Building) of Excellent Engineering Exploration and Design Award of Guangdong Province
  The Second Prize (BIM) of Excellent Engineering Exploration and Design Award of Guangdong Province
  Honorable Mention of the Excellent Architectural Creation Award of Guangdong Province Registered Architect Association
  The Second Prize (Architectural Design - Public Building) of the Guangdong Architectural Design Award of Guangdong Province Registered Architect Association
  Tien-yow Jeme Hometown Cup Prize by Guangdong Society of Civil Engineering and Architecture
  The First Prize of the BIM Application Competition of Guangdong Province
  Sliver Award (Business Category) (gold award left open) of Hong Kong Institute of Architects (HKIA) Cross- Strait Architectural Design Symposium and Awards

- **Zhaoqing New Area Sports Center, Zhaoqing**
  The First Prize of Excellent (Public) Architecture Design of National Excellent Exploration and Design Industry Award
  The Third Prize (Public Building) of ASC Architectural Design Award
  Bronze Award of the Weihai International Architectural Design Grand Prix
  The First Prize (Construction Engineering) of Excellent Engineering Exploration and Design Award of Guangdong Province
  The Second Prize (Green Building) of Excellent Engineering Exploration and Design Award of Guangdong Province
  Tien-yow Jeme Hometown Cup Prize by Guangdong Society of Civil Engineering and Architecture
  Science and Technology Award of Guangdong Society of Civil Engineering and Architecture

- **Satellite Hall, Shenzhen Bao'an International Airport, Shenzhen**
  SKYTRAX "Global Five-star Terminal"
  CAPSE The World's Best Airport
  IAA Muse Design Awards (Platinum)
  Guangdong Green Building Label (Three-star)
  The Second Prize (BIM) of Excellent Engineering Exploration and Design Award of Guangdong Province
  The First Prize (Architectural Design - Public Building) of the Guangdong Architectural Design Award of Guangdong Province Registered Architect Association

- **肇庆东站交通换乘枢纽**
  行业优秀勘察设计奖 · 优秀（公共）建筑设计二等奖
  中国威海国际建筑设计大奖赛优秀奖
  广东省优秀工程勘察设计奖 · 公共建筑一等奖
  广东省注册建筑师协会广东省建筑设计奖 · 建筑方案奖公建类一等奖

- **广东（潭洲）国际会展中心**
  行业优秀勘察设计奖 · 优秀（公共）建筑设计二等奖
  中国会展业金海豚大奖 · 中国会展标志性场馆
  广东省优秀工程勘察设计奖 · 建筑工程二等奖
  广东省优秀工程勘察设计奖 · 全过程咨询（工程总承包）专项二等奖
  广东省土木工程詹天佑故乡杯奖

- **珠海横琴国际科技创新中心**
  亚太房地产大奖 · 中国公共服务建筑优胜奖
  广东省注册建筑师协会广东省建筑设计奖 · 建筑方案奖公建类三等奖

- **湛江吴川国际机场航站楼**
  广东省绿色建筑评价标识 · 二星 A 级绿色建筑设计标识证书
  广东省注册建筑师协会广东省建筑设计奖 · 建筑方案奖公建类二等奖

- **顺德德胜体育中心**
  中国威海国际建筑设计大奖赛优秀奖
  广东省注册建筑师协会广东省建筑设计奖 · 建筑方案奖公建类二等奖

- **南粤古驿道梅岭驿站**
  行业优秀勘察设计奖 · 优秀（公共）建筑设计二等奖
  中国威海国际建筑设计大奖优秀奖
  广东省注册建筑师协会广东省建筑设计奖 · 建筑方案奖公建类一等奖
  广东省优秀工程勘察设计奖 · 传统（岭南）建筑一等奖
  广东省“三师”专业志愿者服务（勘察设计）优秀项目推荐

- **广州猎桥桥西 110kV 变电站**
  电力行业（火电、送变电）优秀工程设计一等奖
  广东省工程勘察设计行业协会科学技术奖一等奖
  广东省土木工程詹天佑故乡杯奖

- **Zhaoqing East Railway Station Transportation Center, Zhaoqing**
  The Second Prize of Excellent (Public) Architecture Design of National Excellent Exploration and Design Industry Award
  Excellence Award of the Weihai International Architectural Design Grand Prix
  The First Prize (Public Building) of Excellent Engineering Exploration and Design Award of Guangdong Province
  The First Prize (Architectural Design - Public Building) of the Guangdong Architectural Design Award by Guangdong Province Registered Architect Association

- **Guangdong (Tanzhou) International Convention and Exhibition Center (Phase I), Foshan**
  The Second Prize of Excellent (Public) Architecture Design of National Excellent Exploration and Design Industry Award
  China Exhibition Industry Golden Dolphin Awards -Landmark Venue of the Year
  The Second Prize (Construction Engineering) of Excellent Engineering Exploration and Design Award of Guangdong Province
  The Second Prize (Whole-process Consulting - EPC) of Excellent Engineering Exploration and Design Award of Guangdong Province
  Tien-yow Jeme Hometown Cup Prize by Guangdong Society of Civil Engineering and Architecture

- **Hengqin International Science and Technology Innovation Center, Zhuhai**
  Asia Pacific Property Awards (Excellence Award of Public Service Building, China)
  The Third Prize (Architectural Design – Public Building) of the Guangdong Architectural Design Award of Guangdong Province Registered Architect Association

- **Terminal of Wuchuan International Airport, Zhanjiang**
  Guangdong Green Building Label (Two-star, Grade A)
  The Second Prize (Architectural Design – Public Building) of the Guangdong Architectural Design Award of Guangdong Province Registered Architect Association

- **Shunde Desheng Sports Center, Foshan**
  Excellence Award of the Weihai International Architectural Design Grand Prix
  The Second Prize (Architectural Design - Public Building) of the Guangdong Architectural Design Award of Guangdong Province Registered Architect Association

- **Meiling Station of South China Historical Trial, Shaoguan**
  The Second Prize of Excellent (Public) Architecture Design of National Excellent Exploration and Design Industry Award
  Excellence Award of the Weihai International Architectural Design Grand Prix
  The First Prize (Architectural Design - Public Building) of the Guangdong Architectural Design Award of Guangdong Province Registered Architect Association
  The First Prize (Traditional Lingnan Architecture Design) of Excellent Engineering Exploration and Design Award of Guangdong Province
  Recommended Excellent Project (Exploration and Design) of Planner, Architect and Engineer Volunteer Association of Guangdong Province

- **110kV Lieqiao Substation, Guangzhou**
  The First Prize (Power Industry - Thermal Power and Power Transmission and Distribution) of Excellent Engineering Design Award
  The First Prize of Science and Technology Award of Guangdong Engineering Exploration & Design Association
  Tien-yow Jeme Hometown Cup Prize by Guangdong Society of Civil Engineering and Architecture

学术论文

1 刘荫培，陈雄 . 高效 · 舒适 · 时代性——广州新白云机场国际竞赛 [J]. 新建筑 ,2001(01):43- 44.

2 陈雄 . 构筑崭新的国际空港——新白云国际机场航站楼设计 [J]. 南方建筑 ,2004(02):75- 78.

3 刘荫培，陈雄 . 广州新白云国际机场航站楼 [J]. 建筑学报 ,2004(09):34- 39.

4 陈雄 . 新白云机场的规划与发展 [J]. 建筑学报 ,2006(07):26- 27.

5 陈雄，潘勇 . 机场航站楼设计的地域性思考——潮汕机场航站楼设计 [J]. 南方建筑 ,2008(01):71- 73.

6 陈雄 . 机场航站楼发展趋势及设计研究 [J]. 建筑学报 ,2008(05):72- 76.

7 潘勇，陈雄 . 广州亚运馆设计 [J]. 建筑学报 ,2010(10):50- 53.

8 潘勇，陈雄 . 广州亚运馆设计与思考 [J]. 建筑创作 ,2010(11):62- 79.

9 刘荫培，陈雄 . 广州新白云国际机场 T1 航站楼 [J]. 建筑创作 ,2010(12):92- 111.

10 陈雄 . 建筑师的摇篮 [J]. 南方建筑 ,2012(05):47.

11 潘勇，陈雄 . 广州亚运馆设计 [J]. 建筑设计管理 ,2013,30(07):30- 34.

12 陈雄，潘勇 . 干线机场航站楼创新实践——潮汕机场航站楼设计 [J]. 建筑学报 ,2014(02):75.

13 潘勇，陈雄 . 广州亚运馆设计 [J]. 建筑设计管理 ,2014,31(07):49- 53.

14 陈雄，潘勇 . 建筑活化 · 新旧无间——广东省院 ADG · 机场院办公楼 [J]. 建筑技艺 ,2014(10):84- 87.

15 陈雄 . 大跨度建筑的形态与空间建构——以机场航站楼与体育场馆为例 [J]. 建筑技艺 ,2016(02):26- 33.

16 陈雄 . 随形共筑：结构成就建筑之美 [J]. 建筑技艺 ,2016(04):23.

17 陈雄，潘勇，赖文辉，郭其轶，易田，钟伟华 . 大型航站楼建筑的多学科一体化设计——以新白云国际机场 T2 航站楼为例 [J]. 城市建筑 ,2017(31):18-21

18 陈雄 . 超大型航站楼设计实践与思考——广州新白云国际机场 T2 航站楼设计 [J]. 建筑技艺 ,2017(12):40- 47.

19 陈雄，陈宇青，许滢，陈超敏 . 传承岭南建筑文化的绿色建筑设计实践与思考 [J]. 建筑技艺 ,2019(01):36- 43.

20 郭胜，陈雄，陈超敏，罗志伟，宋永普 . 肇庆新区体育中心 [J]. 建筑学报 ,2019(05):48- 52.

21 陈超敏，郭胜，陈雄 . 面向可持续性发展的在地营造——肇庆新区体育中心工程实践 [J]. 建筑学报 ,2019(05):53- 55.

22 陈雄，潘勇，周昶 . 新岭南门户机场设计——广州白云国际机场二号航站楼及配套设施工程创作实践 [J]. 建筑学报 ,2019(09):57- 63.

23 陈雄 . 传承地域文化 创作当随时代——当代岭南建筑设计的实践与思考 [J]. 当代建筑 ,2020(01):26- 28.

24 陈雄，郭胜，庞熙镇 . 佛山新福港商住综合体 [J]. 当代建筑 ,2020(05):117- 123+116.

25 郭胜，陈雄，李开建 . 肇庆新区体育中心 [J]. 当代建筑 ,2020(06):60- 67.

26 陈雄 . 湾区机场航站楼设计的加速发展——以广州机场、深圳机场和珠海机场为例 [J]. 世界建筑 ,2020(06):24- 29+144.

27 陈雄 . 粤港澳大湾区交通建筑的城市角色——以白云机场、珠海机场和肇庆东站为例 [J]. 当代建筑 ,2020(10):21- 25.

28 谢少明，陈雄 . 岭南建筑精髓的再诠释——珠海横琴保利中心设计解读 [J]. 世界建筑 ,2020(12): 104- 106+136.

29 陈雄，黄俊华，李珊珊 . 乡土建筑的单元式转译——南粤古驿道梅岭驿站创作实践 [J]. 建筑技艺 ,2020,26(12):48- 55.

30 陈艺然，陈雄，林建康 . 统一建构的设计逻辑——深圳机场卫星厅室内外一体化设计实践 [J]. 建筑技艺 ,2022,28(06):98- 101.

31 林建康，陈雄，陈艺然 . 传承地域文化的绿色建筑设计实践——以深圳宝安国际机场卫星厅为例 [J]. 当代建筑 ,2022(08):17- 21.

32 易田，陈雄 . 枢纽机场航站楼商业区设计比较研究——以广州白云国际机场二号航站楼为例 [J]. 南方建筑 ,2022(11):60- 67.

专著

1 Mark Molen，陈雄．广州新白云国际机场一期航站楼 [M]. 北京：中国建筑工业出版社，2006.
2 陈雄．十年之外 十年之间：广东省建筑设计研究院 ADG 机场设计研究院 2004-2014 作品集 [M]. 北京：中国建筑工业出版社，2014.
3 陈雄，江刚．持守本源 筑梦千里：广东省建筑设计研究院 GDAD65 周年作品集 [M]. 北京：中国建筑工业出版社，2017.
4 陈雄，潘勇，周昶．广州白云国际机场二号航站楼及配套设施 [M]. 北京：中国建筑工业出版社，2020.
5 曾宪川，陈雄，江刚，罗若铭．轨 · 道：交通建筑作品 [M]. 北京：中国建筑工业出版社，2020.
6 冯兴学，陈雄．广州白云国际机场二期扩建工程设计及建设管理实践 [M]. 北京：中国建筑工业出版社，2021.
7 陈加，陈雄，郝晓赛，罗振城．防疫建筑规划设计指南 [M]. 北京：中国建筑工业出版社，2021.
8 陈雄，江刚．臻品营造：广东省建筑设计研究院有限公司 70 周年作品集 [M]. 北京：中国建筑工业出版社，2022.

MILESTONES

1 Liu Yinpei, Chen Xiong. Efficient, Comfortable and Contemporary - International Competition of Guangzhou Baiyun International Airport [J]. New Architecture, 2001(01):43-44.

2 Chen Xiong. Create a Brand New International Airport - Terminal Design of Baiyun International Airport [J]. South Architecture, 2004(02):75-78.

3 Liu Yinpei, Chen Xiong. Terminals of Guangzhou Baiyun International Airport [J]. Architectural Journal, 2004(09):34-39.

4 Chen Xiong. Planning and Development of Baiyun International Airport [J]. Architectural Journal, 2006(07):26-27.

5 Chen Xiong, Pan Yong. Thoughts on Regionalism in Airport Terminal Design - Terminal Design of Chaoshan Airport [J]. South Architecture, 2008(01):71-73.

6 Chen Xiong. Development Trend and Design Research of Airport Terminals [J]. Architectural Journal, 2008(05):72-76.

7 Pan Yong, Chen Xiong. Design of Guangzhou Asian Games Gymnasium [J]. Architectural Journal, 2010(10):50-53.

8 Pan Yong, Chen Xiong. Thoughts on Design of Guangzhou Asian Games Gymnasium [J]. Archicreation, 2010(11):62-79.

9 Liu Yinpei, Chen Xiong. Terminal 1 of Guangzhou Baiyun International Airport [J]. Archicreation, 2010(12):92-111.

10 Chen Xiong. Cradle for Architects [J]. South Architecture, 2012(05):47.

11 Pan Yong, Chen Xiong. Design of Guangzhou Asian Games Gymnasium [J]. Architectural Design Management, 2013,30(07):30-34.

12 Chen Xiong, Pan Yong. Innovative Practice for Terminals of Hub Airports - Terminal Design of Chaoshan Airport [J]. Architectural Journal, 2014(02):75.

13 Pan Yong, Chen Xiong. Design of Guangzhou Asian Games Gymnasium [J]. Architectural Design Management, 2014,31(07):49-53.

14 Chen Xiong, Pan Yong. Building Revitalization and Seamless Integration of the Old and the New - Guangdong Architectural Design and Research Institute·ADG Office Building [J]. Architecture Technique, 2014(10):84-87.

15 Chen Xiong. Shape and Spatial Structure of Large-span Buildings - A Case Study of Airport Terminal and Gymnasium [J]. Architecture Technique, 2016(02):26-33.

16 Chen Xiong. Architecture Based on Structural Form: Structure Makes the Beauty of Architecture [J]. Architecture Technique, 2016(04):23.

17 Chen Xiong, Pan Yong, Lai Wenhui, Guo Qiyi, Yi Tian, Zhong Weihua. Multi-disciplinary Integrated Design of Large-scale Terminal Buildings - A Case Study of Terminal 2 of Baiyun International Airport [J]. Urbanism and Architecture, 2017(31):18-21

18 Chen Xiong. Thoughts on Design of Super Large-scale Terminal Buildings - Design of Terminal 2 of Guangzhou Baiyun International Airport [J]. Architecture Technique, 2017(12):40-47.

19 Chen Xiong, Chen Yuqing, Xu Ying, Chen Chaomin. Thoughts on Green Building Design Practice Combined with Lingnan Architectural Culture [J]. Architecture Technique, 2019(01):36-43.

20 Guo Sheng, Chen Xiong, Chen Chaomin, Luo Zhiwei, Song Yongpu. Zhaoqing New Area Sports Center [J]. Architectural Journal, 2019(05):48-52.

21 Chen Chaomin, Guo Sheng, Chen Xiong. Site-specific Construction Oriented to Sustainable Development – Project Practice of Zhaoqing New Area Sports Center [J]. Architectural Journal, 2019(05):53-55.

22 Chen Xiong, Pan Yong, Zhou Chang. Design for New Gateway Airport in Lingnan Region – Design Practice of Terminal 2 of Guangzhou Baiyun International Airport and Supporting Facilities [J]. Architectural Journal, 2019(09):57-63.

23 Chen Xiong. Carry Forward Regional Culture with Creative Design of Contemporary Spirit – Thoughts on Design of Contemporary Lingnan Architecture [J]. Contemporary Architecture, 2020(01):26-28.

24 Chen Xiong, Guo Sheng, Pang Xizhen. SFK Commercial and Residential Complex in Foshan [J]. Contemporary Architecture, 2020(05):117-123+116.

25 Guo Sheng, Chen Xiong, Li Kaijian. Zhaoqing New Area Sports Center [J]. Contemporary Architecture, 2020(06):60-67.

26 Chen Xiong. Stepped-up Development of Airport Terminal Design in GBA – A Case Study of Guangzhou Baiiyun International Airport, Shenzhen Bao'an International Airport and Zhuhai Jinwan Airport [J]. Journal of World Architecture, 2020(06):24-29+144.

27 Chen Xiong. Urban Role of Transportation Buildings in GBA – A Case Study of Baiyun International Airport, Zhuhai Jinwan Airport and Zhaoqingdong Railway Station [J]. Contemporary Architecture, 2020(10):21-25.

28 Xie Shaoming, Chen Xiong. Re-interpretation of the Essence of Lingnan Architecture – Design Interpretation of Poly Center in Hengqin, Zhuhai [J]. Journal of World Architecture, 2020(12):104-106+136.

29 Chen Xiong, Huang Junhua, Li Shanshan. Unitized Re-interpretation of Rural Buildings – Design Practice of Meiling Station of South China Historical Trial [J]. Architecture Technique, 2020,26(12):48-55.

30 Chen Yiran, Chen Xiong, Lin Jiankang. Design Logics of Consistent Construction – Interior-Exterior-Integrated Design of Shenzhen Airport Satellite Concourse [J]. Architecture Technique, 2022,28(06):98-101.

31 Lin Jiankang, Chen Xiong, Chen Yiran. Green Building Design with Legacy of Regional Culture – A Case Study of Satellite Concourse of Shenzhen Bao'an International Airport [J]. Contemporary Architecture, 2022(08):17-21.

32 Yi Tian, Chen Xiong. A Comparative Study on Concession Area Design of Terminals in Hub Airports – A Case Study of Terminal 2 of Guangzhou Baiyun International Airport [J]. South Architecture, 2022(11):60-67.

MONOGRAPHS

1 Mark Molen, Chen Xiong. Guangzhou New Baiyun International Airport Phase I Terminal [M]. Beijing: China Architecture & Building Press, 2006.

2 Chen Xiong. Within and Beyond a Decade: Collection of GDADRI ADG-Airport Design Group (2004-2014) [M]. Beijing: China Architecture & Building Press, 2014.

3 Chen Xiong, Jiang Gang. Design and Build for Dreams: the 65th Anniversary Collection of GDAD [M]. Beijing: China Architecture & Building Press, 2017.

4 Chen Xiong, Pan Yong, Zhou Chang. Guangzhou Baiyun International Airport Terminal 2 and Supporting Facilities [M]. Beijing: China Architecture & Building Press, 2020.

5 Zeng Xianchuan, Chen Xiong, Jiang Gang, Luo Ruoming. Rail-Way: Collection of Architectural Works (Transportation) [M]. Beijing: China Architecture & Building Press, 2020.

6 Feng Xingxue, Chen Xiong. Guangzhou Baiyun International Airport Terminal 2 and Supporting Facilities Design and Construction Management Practice [M]. Beijing: China Architecture & Building Press, 2021.

7 Chen Jia, Chen Xiong, Hao Xiaosai, Luo Zhencheng. Architectural Planning and Design Guidelines for Epidemic Prevention and Control [M]. China Architecture & Building Press, 2021.

8 Chen Xiong, Jiang Gang. The Road to Excellence: GDAD 70th Anniversary Collection [M]. Beijing: China Architecture & Building Press, 2022.

参考文献

REFERENCES

参考文献

[1] 左丘明 . 国语 [Z] . 春秋时期 .

[2] [古罗马] 维特鲁威 . 建筑十书 [M] . 高履泰，译 . 北京：中国建筑工业出版社，1986.

[3] 国务院规范性文件 . 关于加强设计工作的决定 [Z] . 1956.

[4] [加] 简 · 雅各布斯 . 美国大城市的生与死 [M] . 金衡山，译 . 江苏：译林出版社，2020.

[5] [美] 斯坦 · 艾伦 . 点 + 线：关于城市的图解与设计 [[M] . 任浩，译 . 北京：中国建筑工业出版社，2007.

[6] [美] 斯图尔特 · 布兰德 . 建筑养成记 [M] . 郝晓赛，译 . 北京：中国建筑工业出版社，2019.

[7] 国际现代建筑协会 (C.I.A.M.) 关于城市规划的纲领性文件 [Z] . 雅典宪章，1933.

[8] [丹麦] 扬 · 盖尔 . 交往与空间 [M] . 何人可，译 . 北京：中国建筑工业出版社，2002.

[9] [日] 黑川纪章等 . 新陈代谢宣言：新城市主义策略 [S] .1960.

[10] Charles Jencks.The Architecture of the Jumping Universe [M].John Wiley，1995.

[11] [德] 戈特弗里德 · 森佩尔 . 建筑四要素 [M] . 罗德胤，赵雯雯，包志禹，译 . 北京：中国建筑工业出版社，2010.

[12] [法] 勒 · 柯布西耶 . 走向新建筑 [M] . 杨至德，译 . 南京：江苏凤凰科学技术出版社，2014.

[13] Victor Olgyay. Design with Climate [M] .Princeton：Princeton University Press，1963.

[14] [美] 罗伯特 · 文丘里 . 建筑的矛盾性与复杂性 [M] . 周卜颐，译 . 北京：中国水利水电出版社，知识产权出版社，2006.

REFERENCES

[1] Zuo Qiuming. Discourses of the States [Z]. The Spring and Autumn Period.

[2] [Ancient Rome] Vitruvius Pollio. The Ten Books on Architecture [M]. Translated by Gao Lyutai. Beijing: China Architecture & Building Press, 1986.

[3] Regulatory circular of the State Council of the PRC. Decision on Strengthening Design Practice [Z]. 1956.

[4] [Canada] Jane Jacobs. The Death and Life of Great American Cities [M]. Translated by Jin Hengshan. Jiangsu: Yilin Press, 2020.

[5] [USA] Stan Allen. Points + Lines: Diagrams and Projects for the City [M]. Translated by Ren Hao. Beijing: China Architecture & Building Press, 2007.

[6] [USA] Stewart Brand. How Buildings Learn: What Happens after They're Built [M]. Translated by Hao Xiaosai. Beijing: China Architecture & Building Press, 2019.

[7] Programmatic document released by Congrès International d'Architecture Moderne (CIAM) regarding urban planning [Z]. Athens Charter, 1933.

[8] [Denmark] Jan Gehl. Life between Buildings: Using Public Space [M]. Translated by He Renke. Beijing: China Architecture & Building Press, 2002.

[9] [Japan] Kisho Kurokawa et al. Metabolism: The Proposals for New Urbanism [S]. 1960.

[10] [UK] Charles Jencks. The Architecture of the Jumping Universe [M]. John Wiley, 1995.

[11] [Germany] Gottfried Semper. The Four Elements of Architecture [M]. Translated by Luo Deyin, Zhao Wenwen and Bao Zhiyu. Beijing: China Architecture & Building Press, 2010.

[12] [France] Le Corbusier. Toward an Architecture [M]. Translated by Yang Zhide. Nanjing: Phoenix Science Press, 2014.

[13] [USA] Victor Olgyay. Design with Climate [M]. Princeton: Princeton University Press, 1963.

[14] [USA] Robert Venturi. Complexity and Contradiction in Architecture [M]. Translated by Zhou Buyi. Beijing: China Water & Power Press, Intellectual Property Publishing House, 2006.

跋、致谢

AFTERWORD AND ACKNOWLEDGEMENTS

跋

我与陈雄先生是华南理工大学同专业但不同届的同学，他比我晚一年半年进校，我们同为“文革”后通过高考进入大学的头三届学生，为区别于“文革”爆发时上山下乡的老三届，我们被一些人称为新三届。

建筑师的生涯像一条河，在其发源地，一股股涓涓细流平缓地流淌，经过长时间不动声色地集聚，最终会形成巨大的能量。毕业后的前十多年是陈雄先生潜心工作、努力思考、积蓄力量的时期，最近这20年我常常听到陈雄先生中标和获奖的消息，多年实践经验的积累加上系统思考的习惯使得陈雄先生成为了建筑师中的佼佼者。从陈雄先生和他带领团队的作品中可以明显看出以下几个特点：

关注当代性与地域性

关注地域性是华南理工建筑学院的传统。几十年前学校就成立了亚热带建筑研究室，一代代教授们持续地研究着岭南建筑与园林。陈雄先生的导师林克明教授20世纪20年代毕业于法国里昂建筑工程学院，那时的法国建筑教育已经摆脱了“步扎”系统，林先生回国后尝试了将现代主义精神与地方传统文化相结合的理论与实践之路，在岭南地区留下了大量优秀作品，我将其总结为“追风逐影”，即以“通透”实现自然通风、以遮阳构件创造阴影。陈雄先生延续了前辈将当代性与地域性相结合的理念，其作品不仅具有岭南气候及文化的特色，更是与时俱进，中国改革开放40年来，在经济和技术方面取得了巨大的进步，很多当代新技术在陈雄先生的作品中得到了呈现，同为关注岭南文化，陈雄先生作品不仅体现了地域性，更展现了强烈的时代感。

《国语·郑语》中的“和实生物，同则不继”本是人类发展的辩证法：实现了和谐，则万物生长发育，但若缺少了差异，则会停滞不前。“合”和“和”正是陈雄先生的价值观，陈先生甚至用中国传统造词法创造了一个兼顾形式及内容、其意义可一目了然的新字“[illegible]”，这也太建筑学了！

作品如其人，陈雄先生性格沉稳、为人平和，思维不极端，决策不排他，其建筑作品充分体现出岭南文化承认冲突、兼容并蓄的平和哲学。

从容面对复杂之问题

广州是改革开放的前沿，是中国快速融入全球化时代的典型代表，这里产生了许多具有前瞻性和复杂性的大型公共建筑。作为广州在地建筑师，陈雄先生也积极参与其中，他和团队的许多设计虽然对象还是建筑，但其复杂性已经远远超过了单体，用他自己的话说：这是一座微城市。

在对“功能”的理解上，陈雄先生摆脱了早期现代主义清教徒般的纯粹性和清晰性，面对未来诸多的不确定性，他针对不同建筑类型分别提出了“拆分与重组”“分流改混流”“弹性与预留”等概念，这不仅仅是建筑师的技术思维，更具哲学意味。

建筑师的经验固然重要，但经验毕竟是针对过去之确定问题总结而来，面对未来的诸多不确定性，以建筑师有限的经历所得出的有限的经验未必能百分之百地奏效。陈雄先生带领团队研发和践行由多方参与的、全设计流程的、涉及建筑全生命周期的BIM系统，尝试使用各类软件针对建筑学本体问题进行可量化的建筑空间分析，尤其对于他熟悉的大型体育、交通建筑中人流疏散效率及步行距离的仿真模拟，体现了与时俱进的态度，超越了肉身之人在分析和解决复杂性问题时能力之局限。

高超娴熟的整合能力

从某种意义上说，建筑设计就是处理关系、解决问题、建构秩序，能将建筑设计中诸多矛盾加以巧妙整合是建筑师成熟的标志。

几年前我经过广州白云机场 T2 候机楼时，办票大厅那有节制、又有变化的天花引起了我的注意。在大空间中用变换角度的金属装饰件做吊顶本是建筑师常用的手法，在大空间中部开天窗也是实现自然采光的必然策略，但相当多的大空间建筑中吊顶装饰件仅是装饰而已，我多次见过这种尴尬的情况：由屋面天窗引入的光被密排的装饰构件所遮挡、而装饰构件疏松部位的上空却没开天窗。后来与陈先生当面聊天时才知道，白云 T2 候机楼是他带领广东省建院机场团队原创的，办票大厅上空的铝合金格栅非常巧妙地在天窗处呈现疏松的排列，而在没开天窗的地方格栅逐渐闭合，在解决自然采光的同时也创造了生动的肌理。这是对建筑设计、室内设计与建筑物理的整合。

在亚青会场馆设计时将主场馆降低到 40 米以下，大大降低了巨大体量之建筑对环境造成的尺度压力，还在建筑上开设多种尺度及形态的洞口，并设置多重跌退的平台，这些做法使得建筑与环境之间形成了缓冲界面（interface 或 buffer），这是对建筑、城市与景观的整合。肇庆新区体育中心外立面连续斜向支柱既为建筑增添轻松的气氛，又降低了因风力而产生的侧推力，白云机场 T3 航站楼优雅的组合柱和韵律感十足的天花，深圳宝安国际机场卫星厅的桁架与照明的结合，南粤古驿道梅岭驿站几何形屋面是对传统建筑屋面巧妙的解构与重组，凡此种种，处处体现陈先生和他的团队娴熟的整合技巧和追求完美的哲匠之心！

轻松驾驭尺度及类型

陈雄先生既懂屠龙术，又会杀鸡活儿，其作品中既有超级尺度的城市设计、又有 70 多万平方米的庞然大屋、也有仅区区 500 平方米的袖珍驿站，面对不同尺度的项目，他伸缩自如、游刃有余，既彰显大院总师之熠熠风采，又展示了“螺蛳壳里做道场”的从容不迫。

类型的多样化正是建筑学学科的美妙之处，它为建筑创作提供了无限的想象空间。但作为建筑师中的个体，则很难兼顾全面与深刻，作为超级大院的总师，陈先生和他团队的设计涵盖了大型交通设施、体育设施、科技园区、展览建筑、办公建筑以及住宅等众多类型，且它们都呈现了很高的完成度，这体现了陈先生设计判断及团队把控能力的炉火纯青。在陈先生诸多作品中，有一个几乎不被建筑学列入正册的变电站引起了我的注意：广州猎桥变电站是一个城市中的基础设施，传统意义上的建筑师有可能对其敬而远之，也有可能将其聊做“饭碗”，但陈先生却对其仔细研究，使用多种手法精心设计，使一个冷冰冰的城市基础设施成为了城市人文休闲节点、青少年的科普教育基地和珠江边“月光宝盒”般的新景点。

2016 年陈雄先生被评为全国勘察设计大师，这实乃实至名归。不同于歌星可以一吼成名，建筑师是个相对晚熟的职业，我期待着能持续地看到正处于创作高峰的陈雄先生的新作品问世！

是为跋。

2022 年 12 月 12 日于泰国清迈

李保峰
华中科技大学建筑规划学院教授、博导
国家一级注册建筑师
《新建筑》杂志社长
中国建筑学会绿色建筑学术委员会副主任

AFTERWORD

Mr. Chen Xiong and I were both architecture majors of South China University of Technology (SCUT), while he entered university one year and half after me. We were then called "Xinsanjie" (the "new three classes " of students who started university in 1977-79 after the resumption of Gaokao) by some people. In the first ten years or more after graduation, Mr. Chen worked hard and gained experience. In the recent twenty years, I often heard news about him winning bids and awards. Years of practice, the accumulation of experience and the habit of systematic thinking have made him a prominent architect. The following are the characteristics that clearly mark the works of Mr. Chen and his team:

Focus on modernity and regionalism

There has been a traditional focus on regionalism in the School of Architecture, South China University of Technology, with a State Key Laboratory of Subtropical Building Science set up many decades ago. Following the footsteps of his predecessors, Mr. Chen continued the concept of combining modernity with regionalism, producing works that are not only characterized by Lingnan climate and culture, but also employ many up-to-date new technologies. As noted in Guo Yu·Zheng Yu, "Harmony generates and sameness stifles vitality", which actually explains the dialectics of human development. Upholding "integration" and "harmony", Mr. Chen even created through traditional Chinese word-making a new word " [illegible] ", which has a clear meaning in both form and content. This is so architecture! Like the architect, like his works. Mr. Chen has a calm and peaceful personality, and his architectural design also fully reflects the inclusive and moderate attitude of Lingnan culture towards conflicts.

Simple solutions to tricky problems

Though Mr. Chen and his team are still focused on architecture design, yet the level of the design complexity has far exceeded that of individual buildings, much like designing a micro city, as he said. In his understanding of "functionality", Mr. Chen, in view of different building typologies, has proposed such concepts as "splitting and recombination", "conversion from separate flows to mixed flow", and "elasticity and reservation", reflecting not only architects' professional thinking, but also philosophy, in a certain sense. Mr. Chen and his team have developed and implemented a BIM system involving multi-party participation, the whole design process and the whole life cycle of buildings. Capitalizing on this system, they are experimenting quantifiable analysis of architectural issues with various software, advancing closely with the times.

Strong capacity of integration

In a certain sense, architectural design means to deal with relations, solve problems and establish order. T2 of Guangzhou Baiyun International Airport was the original design of GDAD's airport design team, led by Mr. Chen. The aluminum alloy grating above the check-in hall, which is delicately arranged in a loose pattern at the skylight and gradually narrows elsewhere, addresses the need for natural daylighting while creating a vivid fabric, hence realizing the integration of architectural design, interior design and building physics. In the design of Asian Youth Games Stadium, the main venue is lowered to less than 40 m high to mitigate the suppression of the massive building volume on the surrounding environment. Openings of varied sizes and forms in the building facade, together with cascading terraces of multiple setbacks, function as an interface or buffer between the building and the environment, well integrating the building, the city and the landscape. Other examples include the continuous diagonal columns on the facade of the Zhaoqing New Area Sports Center, which not only add a relaxing feel to the building, but also reduce the side thrust generated by the wind; the elegant colonnade and rhythmic ceiling in T3 of Guangzhou Baiyun International Airport; the combination of truss and lighting in the satellite hall of Shenzhen Bao 'an International Airport; and the geometric roof of Meiling Station of South China Historical Trail, which is an ingenious deconstruction and reorganization of the traditional building roof, etc. These details all reflect the skillful integration and tireless pursuit of perfection by Mr. Chen and his team.

Easy control of scale and typology

Mr. Chen has led his team in delivering design projects of varied scales and typologies, such as the large transportation facilities and sports venues. Among them, Lieqiao Substation, Guangzhou, a substation which has never really been officially recognized as architecture, caught my attention. As part of the urban infrastructure, substation is probably something undesirable for architects in the traditional sense, who would otherwise accept it merely for the sake of remuneration. However, Mr. Chen took it seriously and designed it with various approaches, transforming the once cold urban infrastructure into a cultural leisure node of the city, a popular science education base for teenagers and a new destination like a “Moonlight Treasure Box” by the Pearl River. In 2016, Mr. Chen was awarded the honorary title of National Engineering Survey and Design Master, which he truly deserves. Architects mature slowly and I look forward to more new works from Mr. Chen, who is now in his prime!

End of the afterword.

Li Baofeng
December 12, 2022, in Chiangmai, Thailand

Li Baofeng
Professor and PhD Supervisor of School of Architecture and Urban Planning, Huazhong University of Science and Technology (HUST)
First-class Registered Architect of China
President of New Architecture Deputy Director of Green Building Academic Committee of Architectural Society of China (ASC)

致谢

我想建筑师在设计的一线，总是更多关注项目。今天，这个关于团队多年来设计实践总结的理论成果，能够如期付梓印刷出版，就像完成了一个大项目，这份欣喜也是不言而喻的，心中充满了感恩之情。

我们结合自己的建筑实践，回应建筑创作的核心问题，尝试从设计方法论的角度，形成“合和建筑观”这一理论，希望能在丰富的建筑理论海洋中献一份力。无论是设计作品，还是理论成果，都凝聚了团队的集体智慧。回想在二十多年前，在老一辈专家老总的支持下，我和同事们一起从新白云机场 T1 航站楼开始，至今一直在大型复杂公共建筑的设计领域持续努力。这本专著主要包括了多年追求卓越的 ADG · 机场所的作品，还包括近年成立的朝气蓬勃的城市工作室以及充满创新精神的城规所的作品。这些作品的实现有赖于事业拍档同事伙伴的共同努力和相互信任，有赖于合作各方的大力支持和真诚协作。专著虽然以建筑创作为主线，但需要结构、机电、市政、交通、幕墙、绿建、造价、BIM 等各专业共同支撑建筑创作，才能成就建筑精品。

经过差不多一年的工作，从策划到出版，期间包括主题和大纲的推敲、项目资料收集整理、图书设计排版、文字编写翻译校审，工作非常饱满、计划紧凑而有序推进。在此，衷心感谢各方的大力支持和帮助！

特别感谢马国馨院士应允为本书作序！马院士在交通建筑和体育建筑等领域成就斐然，他的北京首都国际机场 T2 航站楼和国家奥林匹克体育中心是我们学习的标杆。马院士在建筑理论的研究造诣深厚，他的序言令我们这些后辈深受教益，认识到实践与理论紧密结合的重要性。

特别感谢孟建民院士应允为本书作序！正如他在序言所说，交通建筑、体育建筑和医疗建筑一样，都是大型复杂的公共建筑，不断更新迭代，而且迭代速度正在变得越来越快，需要我们不断加强外部协作和学习，孟院士“本原设计”创作观和创作实践是很好的指引。

特别感谢李保峰教授应允为本书写跋！保峰师兄非常熟悉岭南建筑，对传统建筑和当代建构的创新结合深有研究。师兄的“谦和建筑”创作观和创作实践，注重处理人、建筑和自然三者关系，对我们是很好的启发。感谢师兄一直以来的鼓励和帮助！

衷心感谢公司各位领导对本书出版的大力支持，感谢公司相关部门的通力协助！

真诚感谢郭其轶、金少雄、倪俍、陈业文、温云养、黄河清、许尧强、陈超等机场所同事，他们做了大量的具体工作，持续加班加点，大家反复讨论不断推进的场景和氛围，都是温馨的记忆。

感谢各个项目组的同事协助整理资料。本作品集还引用了其他摄影师为部分项目拍摄的照片，在此一并表示感谢！

从 2004 年做西塔竞赛开始，我们的主要出版物都是邀请廖荣辉先生做图书排版和装帧设计，邀请梁玲女士和她的团队做翻译工作，他们不辞劳苦，反复修改调整，追求完美的专业精神一如既往。

今年是广东省建筑设计研究院有限公司成立 70 周年，这无疑是一个非常重要的关键节点。在这个重要的历史时刻，我们需要一个回顾、见证、总结和展望的机会。这本书的出版从一个侧面展示了近二十多年建筑创作的成果，也是一次把实践总结上升到创作理论的新尝试，期待有助于建立创造优秀建筑的工作流程。

借此机会，衷心感谢各界朋友多年来的信任、支持和厚爱！没有你们的大力帮助，我们不可能有机会实现这些作品，不可能有机会取得技术进步。展望未来，对标行业先进，我们期待有更多的好作品出现，这取决于团队的持续努力。好设计，用心做！

陈雄

2022 年 12 月 22 日冬至